PLATFORM

PRESS

Bucks County
Pennsylvania

THE FRACKING TRUTH

AMERICA'S ENERGY REVOLUTION:
THE INSIDE, UNTOLD STORY

CHRIS FAULKNER

For more information, media inquiries, or to contact
the author, please visit: TheFrackingTruthBook.com

Published in the United States by
Platform Press
The nonfiction imprint of
Winans Kuenstler Publishing, LLC
Doylestown, Pennsylvania 18901 USA
www.wkpublishing.com

Platform Press and colophon
are registered trademarks.

Rig site photos courtesy of Breitling Energy
Cover photo: Emanuele Taroni
Front cover design: Tim Evans
Photo of Chris Faulkner: Dan Sellers Photography,
www.DanSellers.com

ISBN: 978-0-9850703-7-3

First Edition

Dedication

This book is dedicated to the nearly 10 million people working in the oil and gas industry, who give their lives, bodies, hearts, and minds to fueling our world. Their ingenuity, grit, and determination are a testament to the industry and an American spirit of fierce independence, boundless optimism, and forward-thinking pragmatism. I am proud and humbled to work alongside the extraordinary people who are creating the American energy revolution.

*"History will be kind to me
for I intend to write it."*
—Winston Churchill

Contents

Figures & Tables

Acknowledgments

Behind every success is a rock. In my case, there are two: my amazing wife Tamra and my devoted mother Carole. You have supported me every day through every adventure, for which my gratitude is boundless. My thanks also to my very talented business partners, Parker and Michael, as well as Mimi and my tremendous group of friends. You have all helped me become who I am today.

About the Author

Chris Faulkner is the President and Chief Executive Officer of Breitling Energy Corporation.

Faulkner founded Breitling in 2004 after designing an advanced exploration software program to identify resource plays that would typically go undetected using traditional methods. He has been a strong advocate of environmentally friendly and socially responsible development of oil and gas resources, creating Breitling's EnviroFrac™ program to evaluate the necessity and environmental safety of additives used in the process of hydraulic fracturing. Breitling has eliminated 25% of the additives used in its frack fluids and strives for 100% reuse or recycling of water used in fracking.

Faulkner was named Oil Executive of the Year in 2013 by the American Energy Research Group, Industry Leader of the Year in 2013 for the Southwest Region Oil & Gas Awards, and was recognized in the *Dallas Business Journal*'s Who's Who in Energy in 2012 and 2013. Under his leadership, Breitling Energy has been named Best Company for Leadership in Oil & Gas Extraction in the US by the IAIR Awards, received the 2013 Aggreko Award for Excellence in Environmental Stewardship from the Gulf Coast Region Oil & Gas Awards, and was named 2013 E&P Company of the Year for the Southwest Region Oil & Gas Awards and Best North American Operator by *World Finance* magazine in 2011, 2012, and 2013.

Breitling Energy's Big Caesar II well drilling ahead.

Preface

I am not a child of the oil business.

Although I grew up in oil country—Texas—I got into the energy field only after I started a company that had developed software that made it possible to find oil and gas in places where it was thought inaccessible or nonexistent. I fell into the energy business by accident and opportunity.

Until about a decade ago, I was a serial entrepreneur, starting and building companies in a variety of industries. I got the bug in junior high when an uncle helped me open a store in my hometown where kids could buy and sell baseball and other trading cards. I started numerous businesses since then—none in the oil industry, until I founded Breitling Oil & Gas Company in 2004 (now Breitling Energy Corporation).

There's no wildcatter in my lineage. No oil company owns me, and no oil company pays me to flog a point of view. I'm a pragmatist, an optimist, and a believer in the concept that capitalism is about more than just the bottom line. The successful business is the one that earns a profit and leaves the communities in which it operates more prosperous and healthy than it found them. I agree with those who say there are more important things in life than making money, but I also know those "more important things" all cost money.

I care just as much about the environment as you. I have a family to whom I want to leave a beautiful and livable world. I was as sick at heart as anyone when the BP oil rig disaster fouled the Gulf of Mexico. I live in a part of the country that experiences prolonged and devastating droughts that some believe may be linked to climate change caused by burning fossil fuels. I travel the world giving speeches in places such as Myanmar and China, where air pollution from coal-fired power plants turns day into night, and those clouds of soot are then carried by the winds across the Pacific to pollute our lakes and streams.

But, as a pragmatist, here's my main point—we can't have it all. We can't expect to have air conditioning at the flick of a switch, or afford-able motor fuels on every corner to keep goods and people moving, or well-lit streets, or high-tech hospitals, or cheap chemicals, or plentiful food, or seamless communications without paying for them, one way or the other. We've been paying for these luxuries by sending our dollars to other countries in exchange for their oil, and in the blood it's cost us to keep that oil flowing. The price has been staggering. At times, it's contributed to nearly wrecking our economy altogether; cost trillions in the direct and indirect expenses of securing the world's major sources of oil; brought death, dismemberment, and devastation to millions on both sides; and helped sow cultural resentments that gave openings to thugs like Osama Bin Laden.

We can't go back, and we can't afford to cross our fingers, hoping that alternative energy sources and conservation will fill the gap before we find ourselves in another war in the Middle East. We can't afford to do nothing until we find ourselves staggering under the burden of gasoline at the $8- to $10-a-gallon prices paid by Europeans, and of natural gas at three to five times what we pay in the US. We've been lucky when you compare us with most of the world. But luck is not an energy strategy for the future.

What I hope to accomplish in these pages is to explain the revolution that's taking place in our own backyards—"fracking"—and is spreading around the globe. There are no silver bullets, but the energy boom has been the catalyst for enormous changes that are just beginning to rein-vigorate our economy, provide us with greater security, give us greater control over our environment, and even help rebuild America's stature in the eyes of the world.

In my frequent travels around the globe speaking to industry and government groups about the fracking phenomenon, it's clear that the world is looking to America—birthplace of the modern energy business—for leadership and technology. Once again, history is blessing the United States. We have the know-how gathered during more than a century, the world needs it, and we're going to sell it to them.

We have discovered or unlocked enormous new sources of energy under our feet. We've made huge strides in conservation and technology, becoming more energy-efficient at a rapid pace. The good old incandescent light bulb is headed for extinction. Smart Wi-Fi thermostats can tell when no one's home and adjust the temperature according to the weather data in your zip code. Energy usage has become an important part of engineering new construction. This phenomenon of energy abundance and efficiency makes it almost a certainty that the cost of powering our nation—already a bargain by international standards—is going to become even less of a burden for our economy for many decades to come.

The revolution has already begun. It's already reducing our oil import dependency, notably from volatile regions or nations antagonistic to America's interests. This revolution may help to defang Russia, which has a clear history of using Europe's dependence on its abundant supply of natural gas as strategic geopolitical leverage. Surging production of natural gas from these new sources is rapidly shrinking the market for coal, widely used by electric utilities and a major source of pollution and greenhouse gases. It's stimulating the manufacture of cars, trucks, and buses powered by natural gas, which burns cleaner and produces far less of the greenhouse gases that contribute to climate change.

In places like China and India, the people I meet share our concern about the planet's health, although you might not think so when you see photos of Beijing at midday that look like dusk. You don't hear much about it in the press at home, but there have been violent protests in a number of Chinese cities over environmental concerns. It's a mistake to think people in other countries value life less than we do. All of humanity, with rare exception, is trying to figure out how to balance the often conflicting goals of stability, prosperity, and a healthy environment.

Every solution to the problem of predictable and affordable energy has risks and a price tag. My aspiration for this book is that readers will come away understanding that the fracking revolution has teed up America to solve this problem with less risk and at greater gain than any other nation on earth. After years of economic devastation and turmoil in our economy, I hope you'll agree that we all have something to work toward and plenty to feel optimistic about.

—Chris Faulkner
Dallas, Texas, 2014

Bakken Shale rig drilling after a heavy snowfall, for Breitling Energy.

Introduction

If you're like most Americans, you've heard and read about fracking, the process of using pressurized water, sand, and trace chemicals to hydraulically fracture rock deep in the earth to unlock new supplies of natural gas, oil, and other hydrocarbons that previously had resisted economic development for decades.

Maybe you've seen one of the films released in recent years that either portray the phenomenon as a major game changer with enormous benefits or as a dire threat to groundwater supplies and the environment in general.

You may have seen news reports of farmers who had been living on subsistence income becoming wealthy after selling rights to energy companies to drill on their lands. Or maybe you've seen film clips of tap water that is so infused with methane gas you can set it afire.

You may have wondered what all the fuss is about. Unless you live in a region that is directly affected by this phenomenon, you're probably more concerned with what's happening in your wallet today than in this brouhaha that seems to have nothing to do with you.

You may not understand how this technology is the catalyst that's changing the world in profound ways, but you probably know within pennies what you paid for a gallon of gas the last time you filled up.

You may not understand the uproar over a proposed oil pipeline from Canada to the US, but you know within a few dollars what your electric bill was last month. If you live in the Northeast and heat with oil, you may not understand why we can't meet all our energy needs with wind and solar power now or in the foreseeable future, but you know how much of an income tax refund you'll need to offset your huge heating bills.

This book is for you if you want to know why the world's leading experts on energy and economics say America is on track to become the largest producer of combined oil and gas on the planet. This book is for you if you want to know what it could mean for you and your children.

This book is for you if you're curious to know how reversing the long decline in America's oil and gas production is not only helping domestic energy security but can help defuse geopolitical tensions for our allies.

This book is for you if you want to know why breakthroughs in the technology we use to find and extract energy from the earth is adding millions of new jobs and is restoring America's role in the world as the leading industrial economy.

With this book, you will learn:

- What America's real energy challenges are, and how all of our energy options carry some economic and environmental trade-offs. Hint: There is no "perfect" energy source.
- How big America's astonishing oil and gas resource base really is—and why we'll never run out of oil and gas.
- What fracking is—and is not. You're in for surprises as to how America's energy revolution came about.
- What is true and untrue about fracking's environmental risks—and why the mainstream media and Hollywood are dominating that message today.
- How the US is already benefiting hugely from the fracking boom, in economic as well as environmental terms—and how it's fueling a new renaissance across other American industries and reviving manufacturing in this country.
- How America will continue to dominate these new sources of energy even as these technology advances spread to other countries with comparable or even greater unconventional resources.
- How fracking may actually prove to be our best near-term solution to the climate change threat without bankrupting the developed world and further impoverishing the developing world.

This book is something of a primer on America's ongoing energy revolution, but it's also a call to action. This is where the oil and gas industry has failed itself and failed the American public. The industry has done a poor job of educating the public on fracking and related technologies that have spawned this revolution. In doing so, it has by default ceded what is the appearance of the moral high ground in this contentious debate to those who would hinder or even ban these technologies, with disastrous results for America.

This book seeks to bridge that information gap and hopefully encourage a healthier dialogue about America's energy choices and their consequences.

Breitling Energy's Pumpkin Ridge #2H drilling ahead.

Chapter 1
America's Energy Challenges

America's Energy Traditions

Long before I joined the oil and gas industry, like many other Americans, I held a somewhat, well, fractured view of the energy business. My perception of it was akin to what you get when you look into a cracked mirror: a distortion created by my own interaction with it having been limited to an immediate, everyday personal level—at the gasoline pump, the home thermostat, the light switch, etc. In short, like most other Americans, I took energy for granted.

Well, that's an American tradition—at least in modern times. This wasn't always the case. If you look at a timeline of this nation's history of energy consumption (Fig. 1-1), you quickly realize that oil and natural gas are fairly recent arrivals as significant sources of energy. And, along with coal, they replaced our reliance on a renewable resource as our primary source of energy: wood. For the first century of our existence as a nation, the typical American family survived by burning wood for warmth and cooking.

Of course, those early settlers also used "slash-and-burn" techniques to clear vast stands of forest for growing crops and grazing livestock. Before you start to get nostalgic about the idyllic, untouched wilderness long preserved by the Native Americans and ruthlessly destroyed by the white man, think again. According to the prominent geography scholar William

Denevan: "The most significant type of environmental change brought about by Pre-Columbian human activity was the modification of vegetation … Vegetation was primarily altered by the clearing of forest and by intentional burning. Natural fires certainly occurred but varied in frequency and strength in different habitats. Anthropogenic fires, for which there is ample documentation, tended to be more frequent but weaker, with a different seasonality than natural fires, and thus had a different type of influence on vegetation. The result of clearing and burning was, in many regions, the conversion of forest to grassland, savanna, scrub, open woodland, and forest with grassy openings."[1]

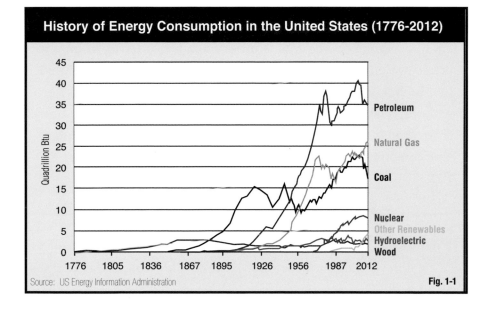

History of Energy Consumption in the United States (1776-2012)

Source: US Energy Information Administration

Fig. 1-1

According to this debunking of what Denevan calls the "Pristine Myth," the mostly European-descended settlers who founded and built the United States of America were just carrying on a tradition that the "real" First Americans had firmly established: using fire as a powerful tool to alter their natural landscape to suit their greater needs and not simply as a primary energy source.

1 http://en.wikipedia.org/wiki/Native_American_use_of_fire

My point here is twofold: 1) Sometimes we make judgments about energy needs versus environmental impacts without knowing all the facts, and 2) often those judgments are based on emotion instead of reason—whether conjuring up myths about eco-conscious aboriginal cultures or wrongfully demonizing an entire industry.

Renewable Energy That Worsens Climate Change?

By the way, wood is indeed a renewable energy source, but it also produces the highest amount of carbon emissions per unit of energy produced—more so than even coal or oil, according to recent research.[1] Unless the burned wood is replaced with new-growth trees (and how fast does that happen, really?), then it's another energy source that is a net negative in terms of greenhouse gas emissions. This is pretty much true of all biomass (solid fuel such as wood and coal) burned to generate heat and power; without offsetting strategies involving new-growth forests and crops that must precede the biomass burn plans, this is one renewable option that can without question worsen climate change.

1 http://www.midwestenergynews.com/2013/05/10/does-burning-wood-instead-of-fossil-fuels-increase-ghg-emissions/

Coal overtook wood as a primary energy source at the dawn of the Industrial Revolution in the late 1880s, and its low cost and easy transportability had much to do with the emergence of American industry. While it took coal a century to supplant wood as a primary energy source in this country, it took only another fifty years or so for oil—and, for a time, natural gas—to overtake coal. But oil products also displaced another well-established energy source: whale oil, used for illumination and lubrication.

I just can't resist noting that the fossil fuels so loathed by environmental pressure groups today helped America reduce deforestation and save the whales.

Since the 1950s, however, coal use rebounded to fuel the hundreds of new electric power plants popping up around the country to serve what seemed at the time a perpetually booming economy. The dawn of the Atomic Age also ushered in another new energy source, nuclear power, which offered the promise of cheap, clean, and almost inexhaustible energy.

But no energy source has so typified America as has petroleum. The oil industry got its start in this country, and its relentless and rapid growth came hand in hand with that other uniquely American-bred industry, automobiles. A prosperous young nation, feeling its oats, spawned what for decades had been the world's biggest and most important industries. Thanks to Henry Ford, the personal automobile was now affordable and widely available. Thanks to Colonel Edwin Drake[2], there was now ample supply of the resource needed to sustain auto industry growth.

And the Seven Sisters—the consortium of major oil companies that controlled over 80% of the world's oil supply—kept oil prices range-bound for decades. From the immediate post-World War II years to 1970, gasoline prices ranged from about 25 cents per gallon to just under 40 cents per gallon. Just the thing for the gas-guzzling behemoths coming out of Detroit. Ads for autos and gasoline were everywhere. No wonder we were complacent.

That complacency was obliterated by the oil price shocks of the 1970s and the birth of environmentalism—ushered in largely by the 1969 Santa Barbara Channel oil spill (the third-largest oil spill in the US, after the 2010 BP Deepwater Horizon and 1989 Exxon Valdez spills)[3] and traffic- and industry-related air pollution concerns. Suddenly, as a nation, we were compelled to worry about energy security and the environmental impacts of our energy choices. But an even bigger worry for the vast majority of American consumers was the spike in gasoline prices at the pump. By 1980, gasoline had topped $1 per gallon for the first time and within a year had reached a level, in inflation-adjusted terms, not far from where it is today. We would never take energy for granted again.

2 http://en.wikipedia.org/wiki/Edwin_Drake
3 http://en.wikipedia.org/wiki/1969_Santa_Barbara_oil_spill

America's New Energy Challenges in a Global Context

The real wake-up call on energy for America was the realization that our country—the birthplace of the modern oil industry and the world's number one oil producer for decades—was not only not number one anymore, but also had become heavily dependent on oil imported from other nations. During the 1970s, both the Soviet Union and Saudi Arabia had surpassed the US in daily oil production; our import dependency was approaching a hitherto unheard-of 50% (Fig. 1-2). The subsequent plunge in US oil imports was the result of two trends: price shock-induced conservation efforts and the start-up of the Trans-Alaska Pipeline System, which brought vast new Alaskan North Slope supplies to the market. US consumption of petroleum had fallen by 3.5 million barrels per day from 1978 to 1983 alone, dropping to its lowest level since 1971.

But a rebounding economy and depressed oil prices had jacked up consumption once again. Oil imports resumed their upward climb, and even as recently as 2004, America's imported oil dependency had rocketed to 60%, and energy analysts were forecasting it would climb even further still well into the new century (Fig. 1-3).

Oil Imports: Share of US Consumption, 1973-2011

47%

Source: US Energy Information Administration

Fig. 1-2

Since Americans have long regarded driving an automobile as part of their birthright, it was inevitable that they would focus their anger at anything that would interfere with that entitlement. Certainly, the Santa Barbara Channel spill was still fresh enough in the public's collective mind to influence that perspective. According to surveys conducted by Opinion Research Corporation, the years following the Santa Barbara Channel spill saw the oil industry's favorability rating decline and unfavorability rating increase.

But those were relatively modest shifts compared to what was to come after the Arab oil embargo of 1973 and the Iranian revolution of 1978–79. Those events spiked oil prices, ushering in the imposition of oil price controls, the creation of the US Department of Energy, and US participation in international emergency oil-sharing agreements. But they also pushed the unfavorability and favorability ratings of the oil and gas industry to record highs and lows, respectively.[4]

It might seem incredible to oil industry observers today, but in the 1960s, the oil industry had a favorability rating near 80% and an unfa-

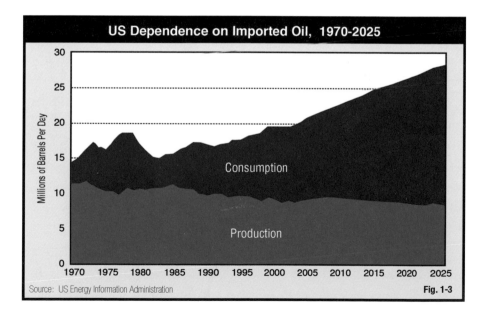

US Dependence on Imported Oil, 1970-2025

Millions of Barrels Per Day

Consumption

Production

Source: US Energy Information Administration

Fig. 1-3

---

4 Bob Williams, *US Petroleum Strategies in the Decade of the Environment* (Tulsa: PennWell Books, 1991), 290-291.

vorability rating in the low single digits. It's doubtful we'll ever get anywhere near those levels again; in fact, recent polls have indicated that the oil industry languishes with tobacco at the bottom of public favorability rankings.[5]

My cynical side might conclude that $4 per gallon gasoline trumps oil spills when it comes to damaging the oil industry's reputation, but then I may as well start asking myself, "Just which circle of hell do you want to occupy?"

Apart from permanently crippling the oil and gas industry's reputation, the oil price spikes of the 1970s created a new awareness of energy security that subsequently dominated our national conversation on energy—at least until the next environmental alarm sounded.

Of course, those energy security concerns largely focused on the perpetual tinderbox that is the Middle East. It is an unfortunate accident of both geology and geography that the lion's share of the world's conventional oil reserves resides in a part of the world that also features some of the world's most vulnerable chokepoints for tankers (Fig. 1-4). More than half of the world's oil production moves to market via tankers on fixed maritime routes. More than one-third of that seaborne trade—or about 20% of all global oil trade—moves through the Strait of Hormuz, a narrow passage at the outlet of the Persian Gulf that separates Iran from the Arabian Peninsula.

Stretching back from the Iran–Iraq war of the 1980s—when the two combatants were lobbing missiles at each other's tankers—to Saddam Hussein's mining the gulf after the start of Desert Storm in 1991, to Iran's recent threats to use mines or scuttle vessels to shut down the Strait of Hormuz in the event of military action related to its development of nuclear weapons, the strait has long been a global energy geopolitics flashpoint.

Is that just bluster from Iran, since a Hormuz shutdown would also cut off most of its own oil exports? It isn't hard to imagine an Iranian leadership that believes in the Twelfth Imam "end of days" scenario and repeatedly vows the destruction of Israel and the United States taking such

---

5 http://www.harrisinteractive.com/NewsRoom/HarrisPolls/tabid/447/mid/1508/
articleId/1349/ctl/ReadCustom%20Default/Default.aspx

a self-destructive path if an actual shooting war breaks out—for instance, if Israel were to destroy Iran's nuclear facilities.

But there are even more vulnerable critical oil supply targets than tankers: the oil and gas pipeline infrastructure in the Middle East (Fig. 1-5). Terrorist bombings of Saudi and other Middle East oil pipelines are a recurring threat, with such attacks escalating in recent years with the emergence of new Al-Qaeda offshoots in the region.

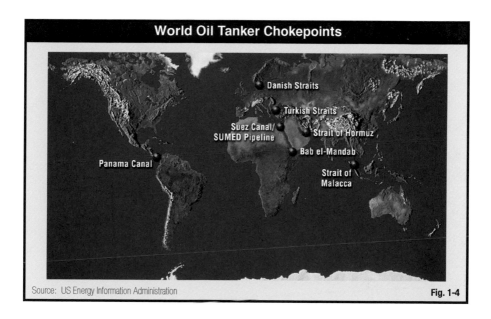

World Oil Tanker Chokepoints

Danish Straits

Turkish Straits

Suez Canal/
SUMED Pipeline

Strait of Hormuz

Bab el-Mandab

Panama Canal

Strait of
Malacca

Source: US Energy Information Administration

Fig. 1-4

It's also important to remember that sudden oil supply outages don't just happen in the Middle East. At this writing, Nigeria and Venezuela, among the world's most important oil producers and OPEC cofounders, are suffering through new periods of violent unrest. And the outages aren't limited to OPEC countries, either. Figure 1-6 shows a persistent state of oil supply outages from non-OPEC and OPEC nations alike during 2011–2013. It's safer to assume that there will be oil supply outages in any given year than otherwise.

Peak Oil & Chicken Little

Not long after the launch of Operation Iraqi Freedom in 2003, we often heard and saw the protest slogan "No blood for oil!" Then, as now, many Americans believe the Gulf wars were prompted, perhaps even designed, to secure Middle East oil fields for the benefit of the major oil companies. With US oil import dependency on the rise and America's emerging role as the world's policeman as the Cold War faded, that scenario took on a greater sense of plausibility for those already predisposed to demonize the oil industry.

About the same time, advocacy of action to combat purported man-made global warming change gathered momentum after emerging as a hot-button issue in the 2000 US presidential election campaign. Once again an environmental issue emerged to cause us to focus intently on our use of oil and gas. And that concern was layered atop energy security worries. It seemed that the only responsible choice was to find clean, secure energy alternatives to oil. Even President George W. Bush, a former oilman himself, decried America's "addiction to oil."

With fresh environmental and energy security concerns, conditions were ripe for a revival of the concept of "peak oil." This concept dates to the mid-1950s, when a prominent—if somewhat eccentric—geologist of his day, M. King Hubbert, postulated that America's oil production would track a type of bell curve, building to a peak before going into a steep downward spiral (Fig. 1-7). Hubbert even predicted that oil production in the US Lower 48 states would peak around 1970 and then enter into decline. And he was right—to a point. But it was a classic case of not seeing the forest for the trees. Here, I'll defer to the most eminent oil historian and scholar of our time, Daniel Yergin, author of the Pulitzer Prize-winning history of the oil industry, *The Prize*:

> *Hubbert used a statistical approach to project the kind of decline curve that one might encounter in some—but not all—oil fields, and he assumed that the US was one giant oil field. His followers have adopted the same approach to assess global supplies.*

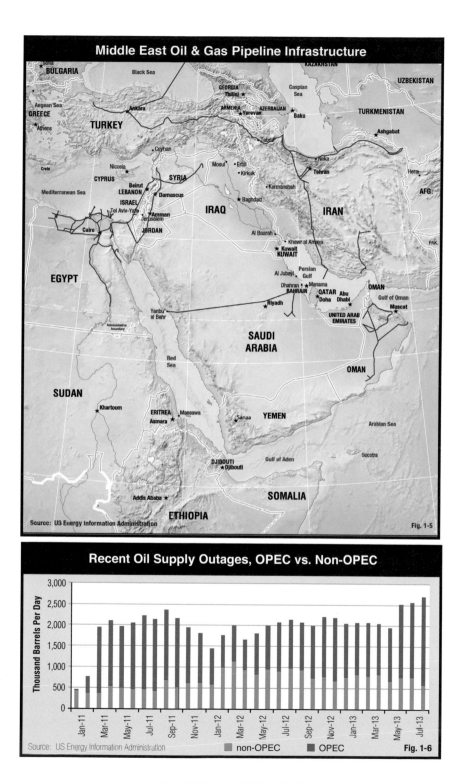

Middle East Oil & Gas Pipeline Infrastructure

Source: US Energy Information Administration

Fig. 1-5

Recent Oil Supply Outages, OPEC vs. Non-OPEC

Source: US Energy Information Administration   non-OPEC   OPEC   Fig. 1-6

Hubbert's original projection for US production was bold and, at least superficially, accurate. His modern-day adherents insist that US output has continued to follow Hubbert's curve with only minor deviations. But it all comes down to how one defines "minor." Hubbert got the date exactly right, but his projection on supply was far off. He greatly underestimated the amount of oil that would be found—and produced—in the US. By 2010, US oil production was 3½ times higher than Hubbert had estimated: 5.5 million barrels per day versus Hubbert's 1971 estimate of no more than 1.5 million barrels per day. Hardly a "minor deviation."[6]

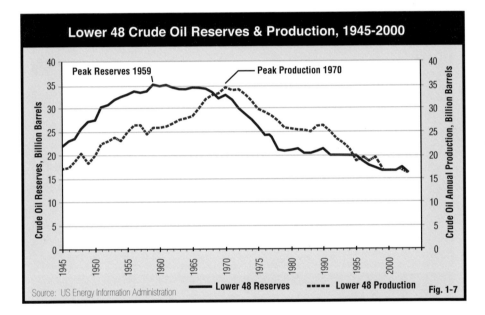

Lower 48 Crude Oil Reserves & Production, 1945-2000

Source: US Energy Information Administration — **Lower 48 Reserves** ----- **Lower 48 Production** Fig. 1-7

Additionally, Hubbert overlooked the contribution to be made by the massive new discoveries on Alaska's North Slope, which did much to slow the nation's oil production decline.

Yergin notes that Hubbert dismissed economics—i.e., the price of oil—and overlooked technology in assuming that oil reserves amount to a static, finite number instead of a moving target that responds to changes in both price and technology. Growing scarcity of a resource increases

6 http://online.wsj.com/news/articles/SB10001424053111904060604576572552998674340

its value and therefore its price. An oil company's existence is defined by its reserves: If it ceases to replace its production with new reserves, it effectively liquidates itself. So an oil company has extremely compelling incentives to innovate new technology to expand its volume of increasingly valuable, enterprise-sustaining reserves.

And while peak oil theory proponents often pointed to the decline in the rate of discovery of new oil reserves in recent decades, they also tended to overlook the fact that most oil reserve growth today comes from extensions of existing oil deposits. In short, the old industry adage, "Oil is where you find it," continues to be validated to this day.

That was acknowledged when the US Department of Energy's statistical and analysis arm, the Energy Information Administration (EIA), got into the act on peak oil. In 2000, the agency developed a dozen scenarios to project the scope and timing of a peak for the world's conventional oil production (Fig. 1-8). At least the EIA acknowledged the variability of price by factoring in demand growth as well as a range of estimates for the ultimate recovery of the resource. That effort yielded peak production year estimates ranging from 2021 to 2112, with likeliest peak year being 2037.

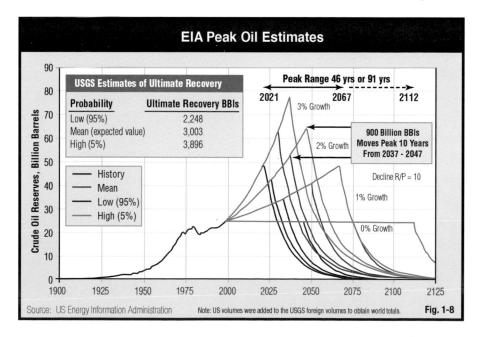

EIA Peak Oil Estimates

USGS Estimates of Ultimate Recovery

Probability	Ultimate Recovery BBls
Low (95%)	2,248
Mean (expected value)	3,003
High (5%)	3,896

Peak Range 46 yrs or 91 yrs
2021   2067   2112
3% Growth

900 Billion BBls
Moves Peak 10 Years
From 2037 - 2047

2% Growth

Decline R/P = 10

1% Growth

0% Growth

History
Mean
Low (95%)
High (5%)

Source: US Energy Information Administration   Note: US volumes were added to the USGS foreign volumes to obtain world totals.   Fig. 1-8

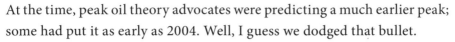

At the time, peak oil theory advocates were predicting a much earlier peak; some had put it as early as 2004. Well, I guess we dodged that bullet.

And all of this analysis and speculation came before the current unconventional oil and gas boom that was spawned by advances in fracking and horizontal drilling. But rest assured that the resource pessimists have regrouped and now are fixated on these newly emerging resources' decline curves as evidence that the current boom is really a bubble. But we'll tell that tale in a later chapter of this book. Maybe we'll call it "Chicken Little: The Sequel."

Access Issues

Okay, so if oil is where you find it, and you know where it is, what do you do when you're not allowed access to it?

Well, during the 2008 presidential election campaign, the Republicans' answer to that question would have been "Drill, baby, drill!" A nation that was worried about growing oil import dependency, a volatile Middle East, and the accompanying increased risk of catastrophic oil spills from increased oil imports (the 2010 BP oil spill in the Gulf of Mexico notwithstanding, the risk of a catastrophic oil spill from a tanker is much greater than that from a drilling rig/platform or a pipeline[7]) should at least be looking in its own backyard for more oil, shouldn't it?

And there is certainly plenty of oil still under our feet in the US. But a lot of that oil (and gas) is off-limits to exploration and development.

According to the American Petroleum Institute (API), at least 87% of our federal offshore acreage is off-limits to drilling. API commissioned the consultancy Wood Mackenzie to assess the foregone offshore opportunity in specific terms. The upshot: Increased access to oil and gas resources underlying federal waters could, by 2025, generate an additional 4 million barrels of oil equivalent per day, add $150 billion to government revenues, and create 530,000 additional jobs.[8] A little

---

7 http://www.boem.gov/uploadedFiles/BOEM/Oil_and_Gas_Energy_Program/Leasing/Five_ Year_Program/2012-2017_Five_Year_Program/Update%20of%20Occurrence%20 Rates%20for%20Offshore%20Oil%20Spills.pdf
8 http://www.api.org/newsroom/upload/soae_wood_mackenzie_access_vs_taxes.pdf

perspective on that first figure: The US consumed about 18.89 million barrels of oil per day in 2013. That's just offshore. The US Department of the Interior's Bureau of Land Management, which oversees America's 279 million acres of public lands, estimated in 2008 that at least 92% of the total 31 billion barrels of oil believed to underlie those public lands is either inaccessible or under restricted-access constraints (Fig. 1-9).

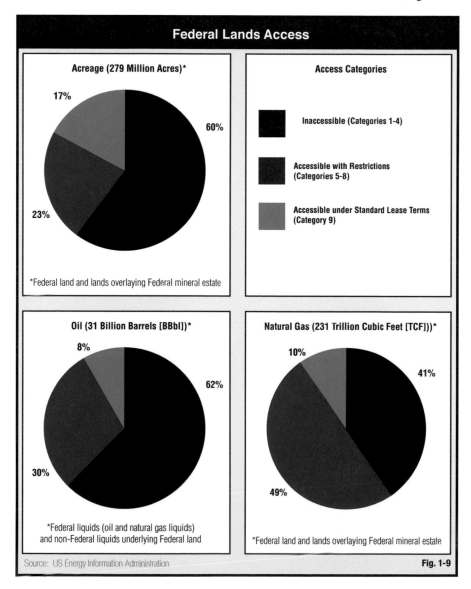

Federal Lands Access

Acreage (279 Million Acres)*

60%
17%
23%

*Federal land and lands overlaying Federal mineral estate

Access Categories

Inaccessible (Categories 1-4)

Accessible with Restrictions (Categories 5-8)

Accessible under Standard Lease Terms (Category 9)

Oil (31 Billion Barrels [BBbl])*

62%
8%
30%

*Federal liquids (oil and natural gas liquids) and non-Federal liquids underlying Federal land

Natural Gas (231 Trillion Cubic Feet [TCF]))*

41%
10%
49%

*Federal land and lands overlaying Federal mineral estate

Source: US Energy Information Administration

Fig. 1-9

For natural gas, the postulated resource is 213 trillion cubic feet (Tcf), of which 51% is restricted or inaccessible. More perspective: The US consumed only 26.03 Tcf of natural gas in 2013.

In fact, since 2007, about 96% of the increase in America's oil and gas production occurred on private lands in the United States. Meanwhile, oil and gas production on federal lands declined to a ten-year low in fiscal years 2011–2012.[9]

And the number of permits issued on federal lands by President Barack Obama is a fraction of his predecessors'—at halfway through his second term, less than one-third the number of federal lands permits issued by the single-term President George H. W. Bush.

I wouldn't expect opponents of drilling on these federal lands to ease up their opposition. If anything, the unconventional oil boom has throttled back the urgency to open up these off-limits lands to exploration and development. However, easing up on that throttle may just prove to be something the oil industry comes to regret.

Efficiency: Low-Hanging Fruit?

Energy efficiency and conservation measures may not jump to top of mind when you think of viable energy resources, but in a very real sense they are key. They don't get nearly as much attention in the energy conversation as does renewable energy. I guess they seem somewhat bland and dull in comparison with gleaming arrays of solar panels and imposing, 130-foot-tall windmills.

But in terms in of having significant impact toward enabling a sustainable energy future, efficiency and conservation might just trump solar and wind. Really? Can flipping a light switch off when you leave the room, knowing when to idle your vehicle rather than shut off the engine, installing weather stripping around your doors and windows, and so on, really make that big a difference in terms of our oil and gas supply/demand balance? Absolutely.

---

9 http://www.eia.gov/analysis/requests/federallands/pdf/eia-federallandsales.pdf

Reducing energy use overall cuts energy costs while it reins in the demand for energy resources. Implementing more energy-efficient technology could even pay for itself if the energy savings offset any additional costs of implementing the technology. Reducing energy use is also seen as a solution to the problem of reducing carbon dioxide (CO_2) emissions. According to the International Energy Agency (IEA), improved energy efficiency in buildings, industrial processes, and transportation could reduce the world's energy needs in 2050 by one-third and help control global emissions of greenhouse gases.[10]

Another way of looking at energy efficiency is to consider energy intensity—energy consumption per unit of gross domestic product (GDP). America has made tremendous strides in this regard. From 1950 to 2011, US energy intensity fell by 58% per real dollar of GDP (Fig. 1-10). But until the 1970s, the decline in energy intensity fell by less than 1% per year. After the 1973 Arab oil embargo-induced oil price spikes, that downward slope got a lot steeper, and the EIA projects energy intensity will continue to decline by about 2% annually through 2040.[11]

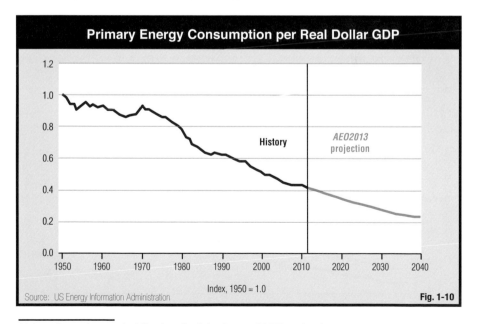

Fig. 1-10

10 http://www.iea.org/publications/insights/name,26319,en.html
11 http://www.eia.gov/todayinenergy/detail.cfm?id=10191

But efficiency and conservation have their limits as well. Investing in a new appliance—say a new electric tankless water heater that may cost six or seven times more than a conventional water heater but promises 20% in energy savings—may sound like a good deal. That is, until you look at the actual life cycle costs and find that the newfangled heater may not recover its additional costs over its lifespan. And sometimes a household budget can accommodate a couple hundred bucks per month indefinitely more easily than it could manage to gin up a couple thousand bucks to buy technology that only promises to make that monthly budget item drop by 20%.

And conserving resources is a nifty idea until it reaches the point where it impedes economic progress. Ever wonder how much money Ebenezer Scrooge was losing over a couple farthings' worth of coal because Bob Cratchit was so cold he couldn't keep a proper ledger? Would China, the world's fastest-growing economy, go on an energy austerity diet if that meant killing its economic growth? Figure 1-11 shows how the world's largest nation has seen its energy consumption grow in lockstep with its GDP.

Additionally, even energy savings from broadly applied conservation and efficiency measures can be overcome by either population growth

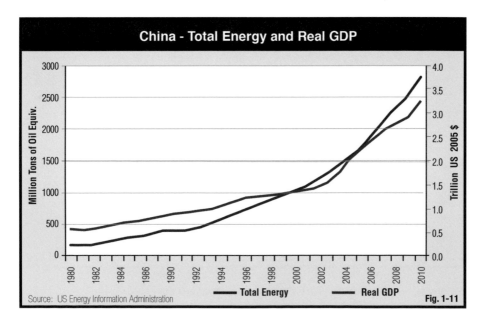

Fig. 1-11

or—and this is really the key to expectations of ever-increasing global energy consumption for decades to come—the transition of a nation from developing to developed status.

Geoff Hiscock, an expert on Asian business, offers this insight about the emerging economic goliath that is China:

> *Every two seconds, somewhere across China, a customer takes delivery of a new car—part of a consumer buying blitz that will see China add 21 million new cars, trucks, and buses to its fleet total in 2014. Short of a catastrophic economic downturn, a government edict against new car ownership, or draconian traffic congestion charges, a continuation of that growth rate means China will likely have a bigger motor vehicle fleet than the United States by 2020. Indeed, the combination of a low vehicle penetration rate—only 85 vehicles for every 1,000 people in China, compared with more than 800 per 1,000 in the US—and the consumer aspirations of high-income, urbanized households across China almost guarantees it.[12]*

In any event, certainly energy efficiency and conservation measures should be considered the low-hanging fruit when it comes to seeking solutions to our energy and environmental challenges. But they don't seem to get promoted as much by the renewables advocates.

Biofuels: From Food to Fuel

Considering that ethanol's origins are tied up with alcoholic beverages, you'd almost think the government planners who mandated the blending of ethanol from corn into gasoline might have had one nip too many of the stuff.

Actually, it didn't seem like such a bad idea in the middle of an oil supply crisis in the late 1970s. A combination of tax credits, tariffs on ethanol imports, and government-backed ethanol plant loans spawned an industry that many at the time thought would bolster our energy security, yield environmental benefits, and help support a depressed agricultural sector.

---

12 http://edition.cnn.com/2013/12/27/business/china-energy-outlook/

Well, they got that last one right, but we're all painfully aware of the associated cost to the rest of us, with the skyrocketing price of corn also spiking up our food costs. According to Aaron Smith, a professor at the University of California-Davis and a specialist in commodity market finance, the competition for corn has created "a vulnerable market in which even the slightest production disturbance will have devastating consequences for the world's poor."

There's been an environmental cost for ethanol production as well, causing long-term damage to our prairie ecosystems and waterways. Since the original ethanol mandates, 1.2 million acres of native grasslands have been plowed under and converted to cornfields—a process that releases carbon dioxide trapped in the soil, increases erosion, and introduces additional chemicals in the form of fertilizers and pesticides into the environment. As much as 5 million acres of land that had been set aside for conservation has been plowed up to plant corn in response to ever-escalating ethanol mandates. There have been countless studies comparing the lifecycle greenhouse gas emissions of producing ethanol from corn with those of gasoline, and the consensus seems to be that, at best, it's a wash.

Then there's the problem of energy content. Ethanol has two-thirds the energy content of gasoline, so it isn't unusual to encounter miles-per-gallon drops of 10–15% with 10% ethanol-blended gasoline.[13] That's an interesting outcome for a fuel designed to be a conservation measure—you have to burn more of it to go the same distance as you could burning less of the fuel you were already using.

From 1979 to 1986, production of ethanol in the US jumped from 20 million gallons per year to 750 million gallons. After more tax credits were applied in the 1990s, ethanol output soared to 3.6 billion gallons per year. The Energy Policy Act of 2005 mandated refiners' blending of 7.5 billion gallons by 2012, and the Renewable Fuel Standard (RFS) law in 2007 hiked it to 15 billion gallons by 2015. In 2007, Congress passed the Energy Independence and Security Act, and ethanol consumption in this country doubled in five years.

---

13 http://www.roadandtrack.com/rt-archive/how-does-ethanol-impact-fuel-efficiency

This latest piece of legislation was premised on forecasts that US oil production would be much lower and oil imports much higher than has been the case, thanks to the fracking boom. The act set ambitious goals for increased biofuel production. At least 37% of the 2011–12 corn crop was to be converted to ethanol and blended with gasoline. It worked, thanks in part to a big, fat government subsidy—producers get a tax credit of up to 45 cents for every gallon they crank out, costing taxpayers about $45 billion.

According to a study[14] by NERA Economic Consulting in March 2013, continued implementation of RFS ethanol mandates by 2015 could:

- Lead to fuel supply disruptions that ripple adversely through the economy
- Cause the cost of diesel to jump by 300% and gasoline by 30%
- Slash the US GDP by $770 billion
- Reduce worker take-home pay by $580 billion

Not the best outcome for Americans still struggling in an anemic economy.

So now we have the unusual scenario of environmental groups on the same side as Big Oil on this issue, along with economists. And consumer organizations have joined in, complaining about the soaring costs at the grocery store. So in November 2013, the EPA surprised refiners by proposing a cutback in the amount of ethanol that's required to be used in motor fuels—from just under 14 billion gallons to about 13 billion gallons.

That's a pretty modest reality check, considering the agency still holds on to its mandate for cellulosic ethanol (from wood chips, corn stalks, or switchgrass) despite the industry's persistent inability to follow the script. In 2013, the production mandate for this corn ethanol alternative was set at 6 million gallons, but only a total of 423,000 gallons was produced. That's a bit of an improvement, considering the original mandate was for 6.6 million gallons, and the yield was … zero.

---

14 http://www.api.org/policy-and-issues/policy-items/alternatives/renewable-fuel-standard-facts

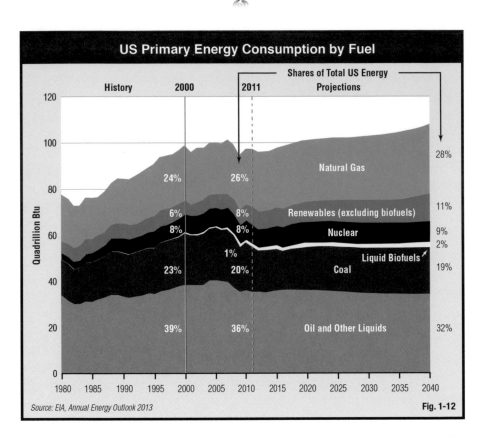

US Primary Energy Consumption by Fuel

History — 2000 — 2011

Shares of Total US Energy — Projections

Natural Gas — 24% — 26% — 28%

Renewables (excluding biofuels) — 6% — 8% — 11%

Nuclear — 8% — 8% — 9%

Liquid Biofuels — 1% — 2%

Coal — 23% — 20% — 19%

Oil and Other Liquids — 39% — 36% — 32%

Quadrillion Btu

Source: EIA, Annual Energy Outlook 2013

Fig. 1-12

Emboldened by that improvement, EPA set the draft mandate for cellulosic in 2014 at 17 million gallons.

By the way, even though the cellulosic ethanol producers have failed to meet the EPA's production target by a very, very long way, gasoline refiners are still held to federal standards for producing mandated gasoline-to-ethanol ratios. The result? Refiners are fined millions of dollars for failing to use the mandated target quantities of a resource that simply isn't there.

So, if ethanol producers could even meet their targets, how much could all that ethanol help our energy security over the long term? Figure 1-12 shows the market share for all liquid biofuels in the US rocketing from the current 1% to … uh, 2% over the next twenty-five years or so. Not exactly an energy magic bullet. And, bear in mind, the chart represents data projected as of early 2013—before the EPA eased up on the ethanol accelerator a bit.

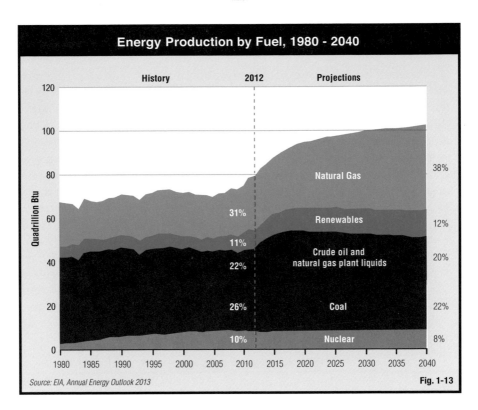

Energy Production by Fuel, 1980 - 2040

History   2012   Projections

Quadrillion Btu

Natural Gas — 38%

31%   Renewables — 12%

11%

Crude oil and
natural gas plant liquids — 20%

22%

Coal — 22%
26%

Nuclear — 8%
10%

1980 1985 1990 1995 2000 2005 2010 2015 2020 2025 2030 2035 2040

Source: EIA, Annual Energy Outlook 2013   Fig. 1-13

Renewables: Always Tomorrow

Long before Al Gore's Nobel- and Oscar-winning documentary *An Inconvenient Truth*, we heard from advocates of renewable energy that wind and solar and other renewable energy sources will save the planet from destruction and the human race from extinction.

But environmentalism really didn't start out focused on "clean" energy so much as other issues. At the first Earth Day in 1970—pretty much Ground Zero for the start of the modern environmental movement—the emphasis was more on air and water pollution, toxic dumps, and wilderness and wildlife preservation.[15] Any complaints about fossil fuels were focused more on oil spills and acid rain from coal-fired power plants.

Flash forward thirty years, when global warming is the new threat and fossil fuels the new enemy. Throughout all of that time the US has endured

15 http://www.earthday.org/earth-day-history-movement

How Hydroelectric Fits into the Energy Puzzle

Hydroelectric power was responsible for about 2.8% of all power generated and about 56% of power generated by renewables in the US in 2012.[1] The US has 5,500 dams, 50 of them considered major. Just the fact that hydroelectric power involves water makes it sound like a clean energy source, and it has, indeed, been touted as such. However, hydroelectric power generation comes with its own environmental baggage. For starters, dam construction displaces a lot of people (an estimated 80 million worldwide)[2] and rich wildlife habitats, destroying entire communities, forests, marshes, agricultural lands, and scenic lands. While water is a renewable resource, the dynamics of dams change the way water evaporates, with higher evaporation rates from dam reservoirs than from the flowing rivers blocked to create them. Dams also impact wildlife on both sides: Reservoirs are more stagnant, typically lower in dissolved oxygen, and a different temperature than the rivers below, killing off many species unable to adapt to the changed conditions of their habitat. Dams restrict the natural flow of sediments and nutrients, so when reservoir water is released

1 http://www.instituteforenergyresearch.org/energy-overview/hydroelectric/
2 http://science.howstuffworks.com/engineering/civil/reasons-dams-happen.htm#page=2

oil supply shortfalls and price shocks and has been embroiled in multiple conflicts in which threats to oil supplies were of paramount concern.

Renewable energy has time and again been touted as a panacea for any geopolitical or environmental threats we face as a nation. Every recent president has committed to supporting renewable energy, opinion polls show overwhelming support for renewables, and the major sources—solar and wind—have made steady progress in reducing their costs to a point where they might soon be competitive with conventional sources of electricity.

So why aren't there solar panels on every home and building? Why aren't the Great Plains covered with windmills? Why aren't we all driving hybrids powered by electricity generated from offshore tidal energy? Why does the EIA project renewables growing their market share of America's energy supply from 11% today to a mere 12% by 2040 (Fig. 1-13)? (Bear in mind, about half of the renewable share is hydroelectric.)

into the river, it can have negative impacts on plant and animal life down-stream. Dams also interfere with fish migratory patterns, although this issue has been addressed in many locations. One of the more surprising environmental impacts of dams is the release of methane and CO_2—not in the operation of the dam but as a result of decaying microorganisms in the stagnant reservoir. While the amount of CO_2 emissions from dam reservoirs (0.5 pounds of CO_2) is smaller than emissions from natural gas-generated (from 0.6 to 2.0) and coal-generated (from 1.4 to 3.6)[3] electricity, the methane generated by the decaying organisms is far scarier: A study by the National Institute for Space Research found that India's 4,500 reservoirs (that's 1,000 fewer than in the US) "emit an amount of methane that is equivalent to 850 million tons of carbon dioxide per year." The study notes that methane is 23 times "more formidable in trapping heat than carbon dioxide."[4] One author, Jacques Leslie, has even reported that, "The world's dams have shifted so much weight that geophysicists believe they have slightly altered the speed of the earth's rotation, the tilt of its axis, and the shape of its gravitational field."[5] Scary baggage.

---

3 http://srren.ipcc-wg3.de/report/IPCC_SRREN_Full_Report.pdf
4 http://science.howstuffworks.com/engineering/civil/reasons-dams-happen.htm#page=1
5 Jacques Leslie, *Deep Water: The Epic Struggle Over Dams, Displaced People, and the Environment* (Farrar, Straus and Giroux, New York, 2005)

ExxonMobil, which projects renewables will grow market share by 150% by 2040, contends that wind, solar, and biofuels will account for just 4% of global energy supply in 2040 versus 1% in 2010.[16]

Is a lack of political will hindering renewables? Conspiracy by the fossil fuel lobbies to crush a promising competitor? Weak marketing efforts?

Could it still just be about the cost? The solar and wind energy that environmentalists promote still can't make the cut economically for a very simple reason: They aren't available all the time, so they need to have some additional backup capacity that keeps working when the wind doesn't blow and the sun doesn't shine.

According to ExxonMobil:

> *Utilities and other power producers around the world can choose from a variety of fuels to make electricity. They typically seek to use energy sources and technologies that enable reliable and*

---

16 ExxonMobil: The Outlook for Energy: A View to 2040 (2014).

relatively low-cost power generation while meeting environmental standards. Over the outlook period, we anticipate that public policies will continue to evolve to place tighter standards and/or higher costs on emissions—including CO_2—while also promoting renewables. As a result, we expect the power sector to adopt combinations of fuels and technologies that reduce emissions but also raise the cost of electricity. At the same time, the sector will also need to manage reliability challenges associated with increasing penetration of intermittent renewables, like wind and solar. These renewables have a cost, which is often overlooked, related to reliability for times when the wind is not blowing and the sun is not shining.

This reliability cost includes added integration into the grid, added transmission capacity, and backup capacity—namely provided by fossil fuel-generated power. ExxonMobil concludes that this reliability cost still puts solar and wind power among the highest-cost electricity sources, even if you factor in the added cost for advanced natural gas- and coal-fired facilities equipped with new technology imposed by carbon limits, whether through carbon taxes or mitigation measures.

What about storing energy from solar and wind technologies for use when sun and wind aren't available? There is a lot of effort going into the development of such storage capabilities, but we do not yet have the ability to cost-effectively store the energy generated by solar and wind technologies, let alone the means to do so in an environmentally friendly way.

A 2013 study by Stanford's Global Climate and Energy Project[17] concluded that "it would actually be more energetically efficient to shut down a wind turbine than to store the surplus electricity it generates." The Stanford study further noted that "Batteries with high energetic cost consume more fossil fuels and therefore release more carbon dioxide over their lifetime. If a battery's energetic cost is too high, its overall contribution to global warming could negate the environmental benefits of the wind or solar farm it was supposed to support."

---

17 http://news.stanford.edu/news/2013/september/curtail-energy-storage-090913.html

Here's another dirty little secret about renewables: They're not so pristine when it comes to environmental impacts or reliable when it comes to energy security.

Hundreds of thousands of birds and bats—species essential for pollination and seed propagation—are pureed each year by windmills.[18] Birds are also getting fried by giant solar installations,[19] as are untold quantities of aquatic insects, which are important food for fish and other aquatic species. Environmental groups themselves are squaring off over the pros and cons of massive new solar and wind facilities placed in sensitive wilderness areas, where they might threaten endangered species.[20]

Concerning the threat to endangered and protected species, consider the examples of two widely disparate energy sources' respective impacts on these birds. The US Department of Justice has aggressively prosecuted oil companies under the Migratory Bird Treaty Act, even bringing criminal charges against oil companies working in the Bakken Shale play of North Dakota when a handful (seven) of ducks and other birds died after landing in the companies' reserve pits. A conviction under the law could result in a $250,000 fine or two years in prison, or both. Now compare that with how the Obama administration is treating the wind industry that is killing scores of bald and golden eagles each year, according to an Associated Press[21] investigation last year:

> Under pressure from the wind-power industry, the Obama administration said ... it will allow [wind-power] companies to kill or injure eagles without the fear of prosecution for up to three decades. The new rule is designed to address environmental consequences that stand in the way of the nation's wind energy rush: the dozens of bald and golden eagles being killed each year by the giant, spinning blades of wind turbines. An investigation

18 http://www.smithsonianmag.com/smart-news/how-many-birds-do-wind-turbines-really-kill-180948154/?no-ist
19 http://online.wsj.com/news/articles/SB10001424052702304703804579379230641329484
20 http://www.businessweek.com/articles/2012-10-04/where-tortoises-and-solar-power-dont-mix
21 http://www.nbcnews.com/business/business-news/wind-farms-can-kill-eagles-without-penalty-f2D11702834

*by the Associated Press earlier this year documented the illegal
killing of eagles around wind farms, the Obama administration's
reluctance to prosecute such cases, and its willingness to help keep
the scope of the eagle deaths secret. President Barack Obama has
championed the pollution-free energy, nearly doubling America's
wind power in his first term as a way to tackle global warming.*

Well, despite that description of wind power being pollution-free,
yes, even solar and wind have CO_2 emissions, as well as incurring
habitat loss, water use, and mining impacts related to producing the
rare earths and other materials used in manufacturing components for
solar arrays and windmills.[22]

Oh, and by the way, China controls 95% of the world's supply of
rare earths, and of course Beijing always acts in a way that benefits
American interests, right? So much for the "energy security" part of the
pitch for renewables.

Nuclear: Promise Unfulfilled

In an ideal world, nuclear energy would meet all of America's electricity
needs. At least that was the promise of what was once seen as a nonpolluting
source of energy that proponents said would ultimately be "too cheap to
meter." (As it turned out, that 1954 description really was intended for ultra-
secret planning for fusion energy, but the perception at the time and since
then was that it applied to nuclear fission.)[23]

Affordable fusion energy—basically how the sun works—is still decades
away, so the mantle has fallen to nuclear power. But nuclear has yet to deliver
on its promise, largely because of grassroots opposition that became especially effective after the Chernobyl disaster (1986) in the former Soviet Union
and the Three Mile Island accident (1979) in Pennsylvania. Until those
events, installed nuclear capacity had grown a hundredfold in the 1970s and
then tripled again in the 1980s. But the specter of nuclear holocaust without
war resulted in nuclear power capacity additions flatlining (Fig. 1-14).

22 http://www.ucsusa.org/clean_energy/our-energy-choices/renewable-energy/
   environmental-impacts-solar-power.html
23 http://en.wikipedia.org/wiki/Too_cheap_to_meter

It's a bit ironic to note that nuclear power, demonized for so long by renewable energy advocates, is actually considered to be a renewable energy itself. The use of uranium-238 in fast breeder reactors, for example, points to a fuel source that effectively could last for 5 billion years. Okay, technically that's a finite number and thus not an infinitely renewable resource, but in 7.6 billion years, the sun, as a red giant, will probably swallow the Earth, anyway.[24] But fast breeder reactors are still prohibitively expensive, and yet they may be a long-term solution for disposing of some nuclear waste, as they burn up some of the waste products that account for long-term radioactivity.

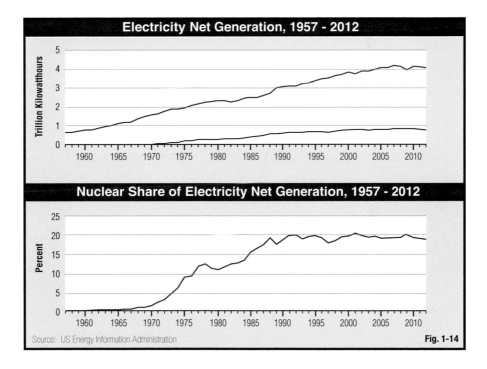

Fig. 1-14

Certainly a case could be made for the great success of nuclear power in France, where the energy source accounts for 80% of its electricity, or even in Japan, where fifty-four nuclear power plants were operating until the horrific earthquake and tsunami of 2011 caused the Fukushima nuclear

---

24 http://www.scientificamerican.com/article/the-sun-will-eventually-engulf-earth-maybe/

plant catastrophe. Up until Fukushima, some renewable energy advocates were starting to rethink their position on nuclear power because of their concerns over climate change. Some experts went so far as to say that cost-effectively reducing carbon emissions to the required level would be impossible without nuclear power.

Even NASA scientist James Hansen, universally acknowledged as the godfather of climate change action advocacy, advocates such a stance. He recently pointed to France's embrace of nuclear power in response to the oil crises of the 1970s as having fostered the fastest decline in greenhouse gas emissions ever recorded.[25] Indeed, the flip side of that development is Germany's two-year increase in greenhouse gas emissions after German Chancellor Angela Merkel shut down the nation's nuclear power plants in the wake of the Fukushima disaster.[26] That was a bit awkward for Merkel, a former environment minister who helped draft the 1997 Kyoto Accord to cut greenhouse gas emissions and combat global warming.

Was Fukushima the last nail in the coffin for nuclear energy? It certainly made the path ahead for nuclear an order of magnitude more difficult, atop the already skyrocketing costs associated with regulatory planning and permitting, litigation, and protracted construction for nuclear power plants. As Deutsche Bank concluded in 2011:

> *The global impact of the Fukushima accident is a fundamental shift in public perception with regard to how a nation prioritizes and values its population's health, safety, security, and natural environ-ment when determining its current and future energy pathways … [consequently] renewable energy will be a clear long-term winner in most energy systems, a conclusion supported by many voter surveys conducted over the past few weeks. At the same time, we consider natural gas to be, at the very least, an important transition fuel, especially in those regions where it is considered secure.[27]*

25 http://www.scientificamerican.com/article/how-nuclear-power-can-stop-global-warming/
26 http://www.bloomberg.com/news/2013-07-28/merkel-s-green-shift-backfires-as-german-pollution-jumps.html
27 Deutsche Bank Group (2011). The 2011 inflection point for energy markets: Health, safety, security and the environment. *DB Climate Change Advisors*

Note Deutsche Bank's descriptives attached to renewable energy and natural gas—"long-term winner" and "important transition fuel." That sounds about right, considering that since Fukushima, Japan's consumption of liquefied natural gas jumped to a 48% market share from 39%, while renewables saw a drop from 10% to 9%.[28]

So, my takeaway is that safe, affordable, clean nuclear energy sounds great. Too bad it's not still 1954 for this energy source.

Coal: Losing Favor Fast

Is it just me, or does most of the climate change activists' ire seem aimed at oil and natural gas instead of coal—which actually produces far more greenhouse gas emissions, especially when compared with natural gas? Not to mention the coal ash produced by power plants burning massive quantities of coal, a nasty byproduct containing high levels of arsenic that most recently threatened the drinking water supply for residents of North Carolina after a defunct Duke Energy plant spilled tons of the toxic stuff into the Dan River in February 2014. This spill followed the massive spill into the tributaries to the Tennessee River from the Kingston Fossil Plant in December 2008, to mention just a few.

Maybe their view is that coal is pretty much already done—as in "stick a fork in it, it's done." That would reflect the growing number of US coal-fired plant retirements in recent years, as well (Table 1-1).

COAL-FIRED GENERATING UNIT RETIREMENTS				
	Existing coal-fired capacity (2012)			
Total net summer capacity (MW)	309,519	1,418	2,456	10,214
Number of units	1,308	29	31	85
Average net summer capacity (MW)	239	49	79	123
Average age at retirement	37	49	58	50
Average tested heat rate (Btu/kWh)	10,168	11,094	10,638	10,353
Capacity Factor	56%	36%	33%	35%
Source: US Energy Information Administration				Table 1-1

---

28 http://www.nbr.org/downloads/pdfs/eta/PES_2013_handout_kihara.pdf

According to the EIA, in 2012 alone, eighty-five coal-fired power plants were retired, or about 3.2% of coal-fired generating capacity, leaving 2013 capacity at 310 gigawatts (GW) spread among 1,308 plants. These retired plants were small and inefficient, as were the sixty plants retired in the preceding two years.

In contrast, units scheduled for retirement over the next ten years are larger and more efficient, according to the agency—at 145 megawatts (MW), 50% larger on average than recent retirements. The agency projects that 60 GW of coal-fired capacity, including that already reported, will be retired by the end of this decade (Fig. 1-15).

To a large degree, coal-fired power plants have been squeezed by sluggish electric power demand but even more so by intense competition from low-priced natural gas spawned by the unconventional gas boom's huge production surge. It's one thing to be the number-one carbon villain when you're the cheapest fuel, but when the competition undercuts you on both price and carbon emissions, it's time to look for a new story line. Of course, the one saving grace for coal in the US has

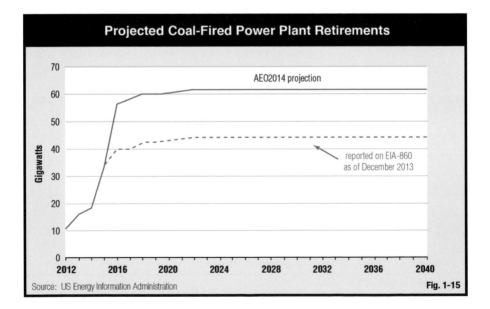

Fig. 1-15

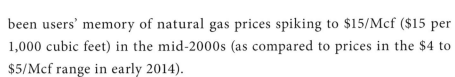

been users' memory of natural gas prices spiking to $15/Mcf ($15 per 1,000 cubic feet) in the mid-2000s (as compared to prices in the $4 to $5/Mcf range in early 2014).

But the emerging story line seems to be one in which the cost of electricity from natural gas remains fairly subdued for the foreseeable future while coal-fired power prices may be spiking—and that's even without new carbon restrictions.

Coal-fired power plants are subject to Mercury and Air Toxics Standards (MATS), which require significant reductions in emissions of mercury, acid gases, and toxic metals. In order to comply with MATS specs, coal-fired plants will have to install flue gas scrubbers or other equipment to remove sulfur and other contaminants. These new standards take effect in April 2015, with a conditional one-year extension option. Accordingly, the EIA reckons that 90% of the aforementioned retirements will happen by 2016.

And this is even before any possible new restrictions related to carbon emissions kick in for coal-fired electricity. While those are largely still hypothetical—and unlikely to kick in while the economy

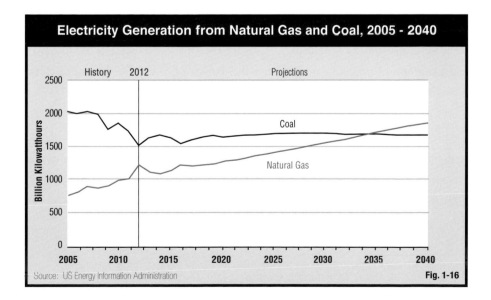

Source: US Energy Information Administration

Fig. 1-16

is still weak—such new strictures undoubtedly provide the impetus for the EIA projecting that natural gas will overtake coal in the electric power market in the mid-2030s, reaching 35% of the market by 2040 versus coal's 32% (Fig. 1-16).

So what options are there for coal producers? Well, one avenue is a technology that captures and stores (carbon capture and storage, or CCS) the CO_2 emitted from a power plant or other industrial facility using fossil fuels. Advanced technology is used to strip the CO_2 from the plant's exhaust system, gather it, and ship it via a pipeline to be injected deep underground.

That sounds appealing at first blush—until you start looking at the economics. According to the US Department of Energy's National Energy Technology Laboratory (NETL), which oversees the government's research on CCS, adding CCS to a conventional pulverized coal power plant spikes the delivered cost of electricity by 80%.[29] For a natural gas-fired power plant, adding CCS jumps the cost by 45%.

But here's an interesting twist: The only commercial track record for injecting CO_2 underground is owned by the oil industry, which has been using natural sources of CO_2 to enhance the recovery of oil from underground formations for decades. (The CO_2 acts as a sort of solvent to make the oil more mobile and sweep it to the wellbore.)

The US leads the world in CO_2 enhanced oil recovery (EOR), mainly with dozens of projects in the Permian Basin of West Texas and eastern New Mexico, some of which have been operating for more than thirty years and have helped oil companies recover billions of barrels of added incremental oil. This technique effectively increases the recovery of the original oil in place in a reservoir from 10–20% under primary recovery to as much as 50–60%. It's been estimated that CO_2 EOR has the potential to effectively double remaining conventional US proved oil reserves to about 45 billion barrels.

---

29 NETL. 2010b. *Cost and Performance Baseline for Fossil Energy Plants Volume 1: Bituminous Coal and Natural Gas to Electricity.*

These CO_2 EOR projects use America's limited natural sources of CO_2, which are costly enough (mainly because of transportation via long pipelines) that the practice hasn't spread much beyond the Permian Basin.

But huge, potential sources of CO_2, of course, are our carbon-belching power and manufacturing/industrial sectors. NETL has been overseeing research into the concept of using America's oil fields to store captured industrial CO_2 emissions by incorporating them into CO_2 EOR projects. This would be a true win-win on energy and the environment, as the oil companies could access affordable CO_2 while providing income to offset the CCS facility cost and using the gas to recover more oil. No doubt tax credits will come into play as well, or some other kind of inevitable government incentives. But some US oil companies are already pursuing such projects without government help, notably Denbury Resources, which focuses on CO_2 EOR projects that include acquiring CO_2 sourced from industrial facilities.[30]

In Canada, a CO_2 EOR project has been operated at Weyburn oil field in southern Saskatchewan since 2000.[31] The project is expected to inject a net 18 million tons of CO_2 and recover an additional 130 million barrels of oil, extending the life of the oil field by twenty-five years. The 95 million cubic feet per day of CO_2 it uses comes from the lignite-fueled Dakota Gasification plant across the border at Beulah, North Dakota (ironically, this originally was the Great Plains coal gasification project that is what remains of the long-defunct, $88 billion Synfuels Corporation boondoggle launched under President Jimmy Carter in 1980). The CO_2 being captured and stored at this project is the equivalent of taking 7 million cars off the road for a year.

The hitch is that America's EOR potential doesn't entirely match up to its potential volume of CO_2 emissions from industrial sources—maybe 10% of the total potential. However, when you factor in the potential for storing CO_2 in depleted oil and gas fields and other underground formations, in addition to EOR, we could be looking at storing all CO_2 emitted from all of our power plants for 100 years.

30 http://www.denbury.com/operations/operations-overview/default.aspx
31 http://en.wikipedia.org/wiki/Weyburn-Midale_Carbon_Dioxide_Project

In the meantime, what are coal producers to do? Turns out they are responding to slack electricity demand and loss of market share to natural gas by exporting more of their coal. US coal exports are booming, and the fastest-growing customer is China.[32]

So America's growing disenchantment with coal results in more coal for the world's biggest emitter of CO_2. If you believe that CO_2 is pollution—personally, I have a little trouble with that, since I exhale it and our ecosystem needs it to survive—then we're just exporting our pollution.

32 http://www.eia.gov/todayinenergy/detail.cfm?id=11791

Breitling Energy's Big Horn #1H well drilling in winter amid baled hay.

Chapter 2
America's Oil & Gas Bounty

More Oil & Gas Than We'll Ever Use

The US will never run out of oil and natural gas. Never. Same goes for the rest of the world. Won't happen.

Whoa, you say, aren't oil and gas finite, nonrenewable resources? Doesn't that mean that we will inevitably run out?

Well, I could dredge up an old oil and gas industry debate over whether oil and gas are biogenic or abiogenic in origin. Most petroleum scientists believe oil and natural gas are biogenic—formed from the compression of the remains of organisms over eons (hence the popular notion that our oil and gas resources come from the decomposed remains of the dinosaurs). But a minority view holds that oil and gas are formed from non-biological processes that convert carbon from the Earth's mantle into lighter carbon compounds that percolate upward to the Earth's surface as oil and natural gas. This latter view would mean hydrocarbons are constantly being formed everywhere—and hence are renewable (and infinite, for all practical purposes)—instead of being geologically and geographically restricted and nonrenewable. In other words, oil and gas would no longer be considered finite fossil fuels.

But that esoteric debate isn't the point here. The total aggregated volume of America's hydrocarbon endowment—or the world's, for that matter—is so vast that, under the most aggressive consumption scenario, we couldn't possibly consume it all.

Methane Hydrates: Fire Ice

Take, for example, methane hydrates, also known as methane clathrates, or "fire ice."[33] These are cage-like lattices of frozen water molecules that contain methane—the main component of natural gas—and can occur below the arctic permafrost and beneath the ocean floor around the world (Fig. 2-1). They can also be a dangerous nuisance when they form spontaneously during deepwater drilling and production operations.

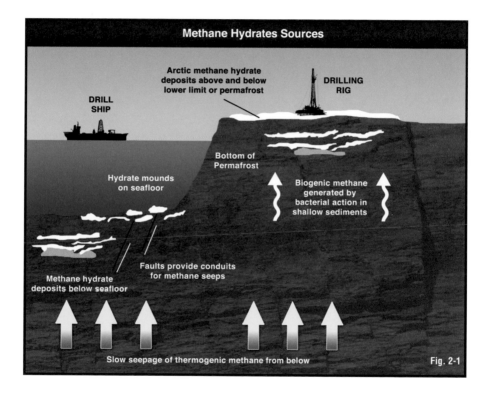

Fig. 2-1

Here's the kicker: Methane hydrates may contain more organic carbon than all the coal, oil, and other forms of natural gas combined, according to the US Geological Survey. The US Department of Energy's National Energy Technology Laboratory (DOE/NETL), which oversees federally funded research of this little-known but vast resource, notes that estimates

---

33 http://www.netl.doe.gov/research/oil-and-gas/methane-hydrates

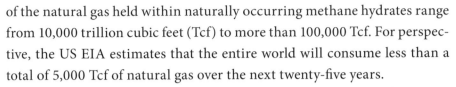

of the natural gas held within naturally occurring methane hydrates range from 10,000 trillion cubic feet (Tcf) to more than 100,000 Tcf. For perspective, the US EIA estimates that the entire world will consume less than a total of 5,000 Tcf of natural gas over the next twenty-five years.

Producing natural gas from methane hydrates isn't theoretical. A DOE/ NETL-funded test project produced a steady flow of gas from a methane hydrate source on Alaska's North Slope in 2012. But the methane hydrate resource remains non-commercial, and with the shale gas revolution, it's hard to imagine why anyone would try to pursue commerciality of methane hydrates production when there is such an abundant source of natural gas already available and commercially feasible. That brings us back to the red herring that is peak oil and gas. (Yes, natural gas was once part of that debate too, but you don't hear much about peak gas anymore, given the unexpected shale gas boom.)

That's the basic point here. To recall an industry adage first coined by Saudi Oil Minister Sheikh Ahmed Zaki Yamani in the 1970s, "The Stone Age didn't end for lack of stones, and the oil age will end long before the world runs out of oil."

In other words: The world's total hydrocarbon resource is more than enough to accommodate the world's energy needs almost indefinitely, assuming sufficient capital, technological innovation, and political will can be thrown at it. We will certainly develop better, cleaner, and even more affordable energy alternatives before we even get close to the point where ultra-costly and ultra-complicated oil and gas resources such as methane hydrates are needed.

But for the next several decades, the best energy choices from a basic dollars-per-unit-of-energy-output model will continue to be fossil fuels. That's why all serious forecasts show fossil fuels dominating global energy consumption—typically accounting for 75% of the total—to the midpoint of the century (Fig. 2-2). Put simply, fossil fuels are the most affordable and efficient energy choices for the foreseeable future, especially for developing nations, and if the low-cost versions (conventional oil and gas) start to get a bit scarce, then price and technology join up to develop new, if

slightly more costly, versions to fill the gap. Sometimes it can be a big breakthrough, such as the marriage of advanced horizontal drilling efficiencies to the enhanced capabilities of multistage hydraulic fracturing that spawned the current unconventional resource boom. Other times it can be the accumulation of many small improvements to drive reserves growth—a hugely overlooked component of oil and gas supply.

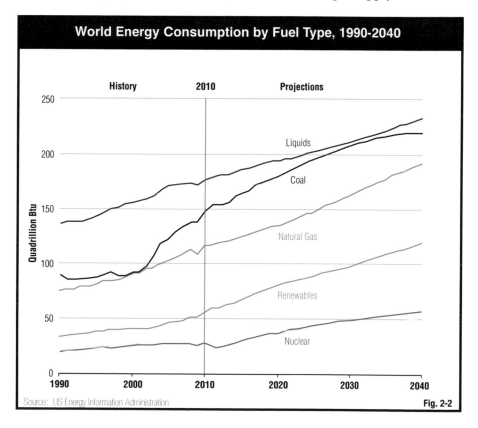

World Energy Consumption by Fuel Type, 1990-2040

Fig. 2-2

Source: US Energy Information Administration

Reserves Growth: A Humble, Unsung Hero

This is a broader view of what agitates the peak oil crowd so much, as we saw in the first chapter. If you take a static view of a fixed number for oil reserves, of course it will follow a bell curve track: Production builds to a certain midpoint and then irretrievably declines—but only if that original fixed number never changes. However, that reserves number is a moving target, and it changes in response to incentives such as price, access, or technology. Any of those factors can result in another portion of the underlying resource being converted to reserves.

Clarifying Reserves Versus Resources

Here's a quick note on terminology that the mainstream media often get wrong when they try to parse the significance of numbers tied to oil and gas potential. When industry experts talk about oil and gas reserves, they are usually talking about proved reserves. There are various categories of reserves—proved, probable, and possible—that typically mean oil and gas that is technically and economically recoverable depending on current prices; the cost of drilling, completing, and producing wells; and the ultimate recoverable volume of oil or gas from those wells. Resources, on the other hand, are oil and gas volumes that could be produced with current technology without any consideration for prices or costs. Huge difference. Resources are a nice, pie-in-the-sky kind of number. But reserves can be booked as assets. A commodity price collapse, such as we saw most recently in 2008–2009, can dramatically slash the value of those assets. Billion-dollar write-downs are not unusual in the oil and gas industry when commodity prices collapse. Because they can be booked as assets, the US Securities and Exchange Commission (SEC) tends to look askance at operating companies that misreport the status of their oil and gas resources/reserves, and will put those valuations to a stern test. Reserves are classified as proven (1P or P90, for a 90% likelihood of economic recovery), probable (2P or P50, for 50% likelihood), and possible (3P or P10, for 10% likelihood). Until January 2010, the SEC didn't allow US oil companies to report anything other than 1P, or proven, reserves, to their investors. Even though that roadblock has since fallen, most US oil and gas operating companies still just report their 1P reserves.

This explains the paradox of why the worldwide oil and gas reserves total continues to grow even as the total volume of prospective reserves from oil and gas discoveries fails to offset the loss of reserves from production each year.

Figure 2-3 shows how US oil production declined steadily since the early 1980s, when the Trans-Alaska Pipeline carrying production from supergiant Prudhoe Bay oil field reached full capacity. Yet oil discoveries lagged so badly that even this reduced level of production should have depleted our oil reserves entirely a long time ago—if all you counted as reserves growth were discoveries. The total volume of oil produced in that time span was seven times the additions to reserves from discoveries. Instead, revisions and extensions to reserves in existing fields made up most of the difference.

Note the increases in all three chart categories after 2007, as reported

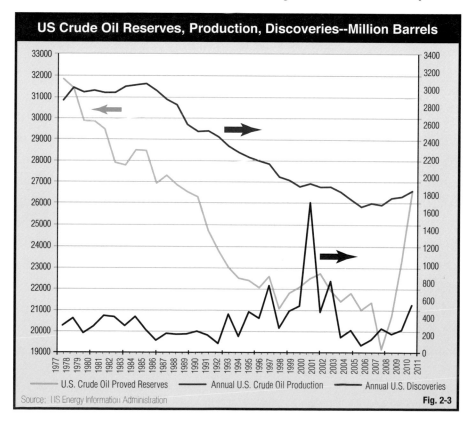

US Crude Oil Reserves, Production, Discoveries--Million Barrels

—— U.S. Crude Oil Proved Reserves —— Annual U.S. Crude Oil Production —— Annual U.S. Discoveries

Source: US Energy Information Administration

Fig. 2-3

by EIA, reflecting the onset of the boom in shale and other unconventional resources. (The big spike in discovered reserves in 2001 reflected the start-up of giant fields in the deepwater Gulf of Mexico.)

What Figure 2-3 doesn't show is the all-time peak for US oil reserves, 39 billion barrels in 1970—a jump of 10 billion barrels in just a year with the Prudhoe Bay discovery. More recent data from *Oil & Gas Journal* shows the US has rebuilt its proved crude oil reserves to almost 32 billion barrels as of the first of this year, a whopping 68% increase since the recent-era low point of 19 billion barrels in 2008.[34] The reserves added through discovery in that time was a fraction of that gain, with most of the increase coming from the giant oil shale plays' additions to reserves.

That 10-billion-barrel increase in reserves in 1970 was essentially all Prudhoe Bay and reflected the original estimate of what portion of that mammoth field's 25 billion barrels of original oil in place could be recovered. After producing 12 billion barrels over the past thirty-six years, it is now estimated that Prudhoe could yield yet another 4 billion barrels of oil—a 60% increase in proved reserves from the original estimate. Like I said, a reserves number is a moving target.

America's Big Endowment

Proven reserves are only a small subset of America's total endowment of hydrocarbon resources. Figure 2-4 shows a snapshot of just the estimated conventional oil resources in the US. It shows that, of the total 627 billion barrels of the technically recoverable discovered oil in America, less than 200 billion barrels has been produced, 407 billion barrels remains to be produced, and more than half of that remainder lies at shallow depths. Oilmen would regard this remaining volume as "stuck behind pipe" (oil that is technically producible but not at the moment for lack of price, market, or takeaway capacity).

But this pie chart created by DOE/NETL reflects data way back in 2002, long before the unconventional oil boom really started to gather momentum.

---

34 http://www.ogj.com/articles/print/volume-111/issue-12/special-report-worldwide-report/worldwide-reserves-oil-production-post-modest-rise.html

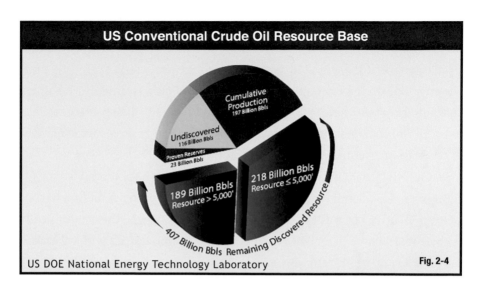

US Conventional Crude Oil Resource Base

Cumulative Production
197 Billion Bbls

Undiscovered
116 Billion Bbls

Proven Reserves
23 Billion Bbls

189 Billion Bbls
Resource > 5,000'

218 Billion Bbls
Resource ≤ 5,000'

407 Billion Bbls Remaining Discovered Resource

US DOE National Energy Technology Laboratory

Fig. 2-4

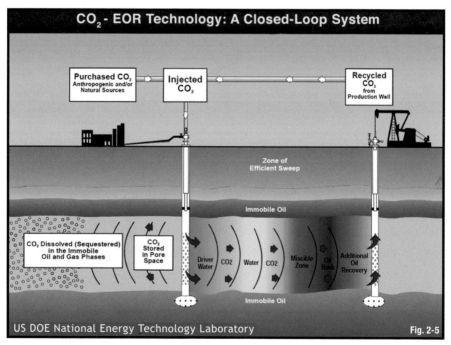

CO_2 - EOR Technology: A Closed-Loop System

Purchased CO_2
Anthropogenic and/or
Natural Sources

Injected
CO_2

Recycled
CO_2
from
Production Well

Zone of
Efficient Sweep

Immobile Oil

CO_2 Dissolved (Sequestered)
in the Immobile
Oil and Gas Phases

CO_2
Stored
in Pore
Space

Driver
Water

CO2

Water

CO2

Miscible
Zone

Oil
Bank

Additional
Oil
Recovery

Immobile Oil

US DOE National Energy Technology Laboratory

Fig. 2-5

NETL created this pie chart in support of its efforts to advance research in CO_2 EOR, which we described in Chapter 1, and that is still being considered as a viable option to reduce greenhouse gas emissions as well as boost oil recovery. Figure 2-5 is a schematic of how such a CO_2

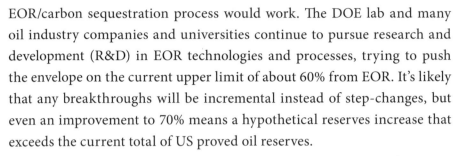

EOR/carbon sequestration process would work. The DOE lab and many oil industry companies and universities continue to pursue research and development (R&D) in EOR technologies and processes, trying to push the envelope on the current upper limit of about 60% from EOR. It's likely that any breakthroughs will be incremental instead of step-changes, but even an improvement to 70% means a hypothetical reserves increase that exceeds the current total of US proved oil reserves.

DOE/NETL has conducted extensive modeling of the potential from CO_2 EOR in the US, identifying 1,858 large oil reservoirs with 366 billion barrels of original oil in place as favorable for CO_2 EOR. It pegged the economically recoverable volume of oil from CO_2 EOR at 67 billion barrels with oil at $85/barrel, CO_2 at $40/metric ton, and a 20% before-tax return.[35] That could support added production of 4 million barrels per day compared with the current total US EOR output of about 780,000 barrels per day.

CO_2 EOR isn't the only EOR process actively contributing to America's oil supply. Massive steamflood EOR projects in California's giant heavy oil fields have been operating for decades. Together, these two processes account for almost all EOR oil production in the US, with CO_2 EOR accounting for about 60% of total output versus steam and other thermal methods with accounting for about 39%.[36]

But we're still just talking about conventional oil reserves. When you look at all potential sources of oil, the numbers are staggering. Consider kerogen shale oil. This gets a bit confusing, so bear with me. Kerogen shale oil is not the same thing as crude oil from shale formations such as the Bakken or Eagle Ford shales. It's a mudstone that holds a lot of an organic sedimentary material called kerogen—which is basically a precursor to crude oil (in essence, it's oil that hasn't been "cooked" enough).

America has about two-thirds of all the kerogen shale oil in the world, with a total oil resource in place of about 6 trillion barrels, concentrated mainly in the Rocky Mountains region. It's been estimated that as much

35 http://www.netl.doe.gov/File%20Library/Research/Energy%20Analysis/Publications/
DOE-NETL-2011-1504-NextGen_CO_2_EOR_06142011.pdf
36 http://www.ogj.com/articles/print/volume-112/issue-4/special-report-eor-heavy-oil-survey/
survey-miscible-co-sub-2-sub-continues-to-eclipse-steam-in-us-eor-production.html

as 800 billion barrels of that volume could be recoverable and bookable as reserves.[37] That dwarfs the proved reserves of Saudi Arabia. Trouble is, no one has really cracked the code for getting oil out of kerogen shale yet. Billions were spent by industry and government in the early 1980s in an attempt at a Rockies kerogen shale boom that went bust after oil prices collapsed and no major breakthroughs were accomplished.

Then we have tar sands, also known as oil sands. This is a combination of clay, sand, water, and bitumen (a heavy, viscous oil) that is either mined or recovered by injecting steam or another heat source underground. Canada, of course, has the world's largest tar/oil sands resource and really the only significant commercial oil sands industry.

In the US, the tar sands resource lies mainly in eastern Utah with an upside estimate of the oil resource in place at 19 billion barrels. Again, this is a resource that has not been commercially produced in the US despite its huge potential. Unfortunately, the Canadian oil sands experience can't be replicated here because the Utah tar sands have less of an affinity for water at the molecular level, which makes the oil less mobile and much more difficult to produce.

Shale/Tight Oil & Gas: The Game-Changer

All of these aforementioned unconventional oil and gas resources have yet to move the needle on US oil and gas production, and even the conventional-oil solution of CO_2 EOR remains constrained by a current lack of natural CO_2 sources.

There are other massive unconventional oil and gas resources in the US that already are moving the energy supply needle in a big way, and how and why that has happened, and will continue despite much controversy and misinformation, is pretty much what prompted me to write this book.

Some readers will immediately jump to the conclusion that I'm talking about shale oil and gas, but the discussion here goes beyond that in terms of unconventional resources. And here we have another issue with confusing

37 http://energy.gov/sites/prod/files/2013/04/f0/Research_Project_Profiles_Book2011.pdf

terminology. It's difficult enough to deal with the confusion between shale oil and kerogen shale. A lot of folks throw the word "shale" around recklessly, applying the label to just about any play that requires hydraulic fracturing to be economic or even to maximize returns. But it's often a misnomer (see sidebar), and more importantly, limiting the discussion of America's current oil and gas revolution to any one resource misses the point about the truly revolutionary nature of what is happening in the industry today. We'll pick up this thread of the conversation in the following chapter, but for now let's focus on the overall resource potential itself.

Starting on the gas side, the three major active, producing sources of unconventional gas in the US are shown in their typical reservoir settings in Figure 2-6. We'll tackle the smallest and least economic of the three major unconventional gas resources first.

A Note on Definitions

Shale, among the most common of rocks, is a finely grained, laminated sedimentary rock made up of clay flakes and bits of other minerals. It has a high content of organic matter that is broken down into gas or oil by high temperatures from compression of other rock layers over eons. It is extremely dense material with very low permeability (think of a brick versus a sponge). Tight sand, on the other hand, has a comparably low permeability, but entails largely sandstone reservoirs. Tight oil and gas are found in limestone and carbonate formations as well. A major distinction is that shale has always been regarded as mainly a source rock for shallower oil and gas reservoirs but can be an oil or gas reservoir when certain characteristics are present. A third major unconventional gas source is coalbed methane, also known as coal seam gas, in which the methane is adsorbed (molecules or particles bound to a surface, as opposed to dissolved or permeated when absorbed) into the coal matrix itself.

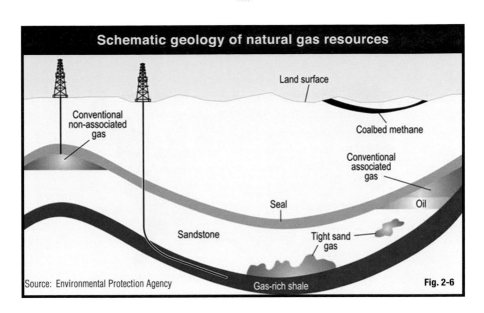

Schematic geology of natural gas resources

Source: Environmental Protection Agency

Fig. 2-6

Coalbed methane (CBM) is recovered via vertical or horizontal wells drilled into coal seams that are too deep or thin to be commercially mined for coal. The methane is recovered by pumping water into the seam, which causes the gas to be desorbed (released from a surface) and the methane is produced along with the water. It has been a significant source of natural gas production for decades but recently fell into a slump because of low gas prices and environmental concerns (handling produced water, regulatory restrictions). CBM resources are estimated at 700 Tcf in the Lower 48 states and 1,000 Tcf in Alaska, while the Lower 48 recoverable volumes are estimated at 100 Tcf (Fig. 2-7).[38]

Until recently, tight sands gas was the dominant contributor to US gas supply among the unconventional gas resources and today still accounts for about 35–40% of all unconventional gas production—mostly from the Rocky Mountains region (Fig. 2-8). Tight gas sands development typically entails directional drilling (I'll spare you the technical distinctions here and just say that any well that is angled off from the vertical but that is not strictly horizontal 90° angle is a directional well) combined with either hydraulic fracturing or acidizing treatments. Generally speaking, while tight sands wells

38 http://www.netl.doe.gov/publications/factsheets/policy/Policy019.pdf

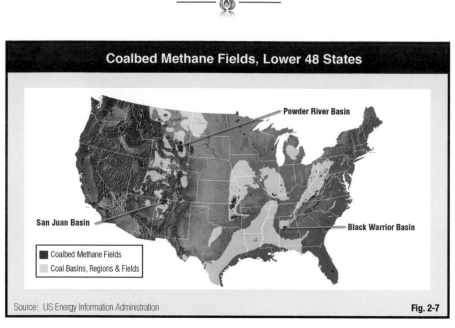

Coalbed Methane Fields, Lower 48 States

Powder River Basin

San Juan Basin

Black Warrior Basin

■ Coalbed Methane Fields
☐ Coal Basins, Regions & Fields

Source: US Energy Information Administration

Fig. 2-7

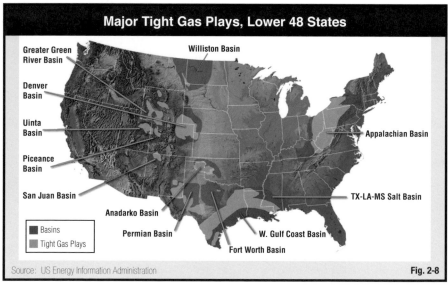

Major Tight Gas Plays, Lower 48 States

Greater Green River Basin

Williston Basin

Denver Basin

Uinta Basin

Appalachian Basin

Piceance Basin

San Juan Basin

TX-LA-MS Salt Basin

Anadarko Basin

■ Basins
■ Tight Gas Plays

Permian Basin

W. Gulf Coast Basin

Fort Worth Basin

Source: US Energy Information Administration

Fig. 2-8

cost less to drill and complete than do shale wells, they are not as prolific. Recoverable tight sands gas resources in the US are estimated at 310 Tcf.[39]

I'll digress from the unconventional gas discussion for a moment to discuss the category of tight oil sands. This is basically really just

39 http://www.ogfj.com/unconventional/tight-gas.html

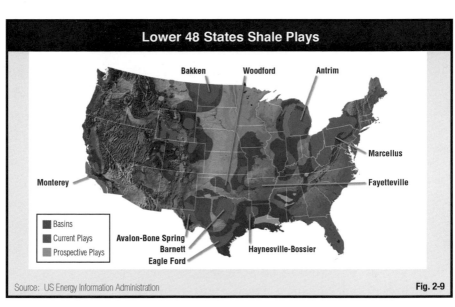

Lower 48 States Shale Plays

Bakken   Woodford   Antrim

Marcellus

Monterey   Fayetteville

Basins
Current Plays   **Avalon-Bone Spring**
Prospective Plays   **Barnett**   **Haynesville-Bossier**
                     Eagle Ford

Source: US Energy Information Administration

Fig. 2-9

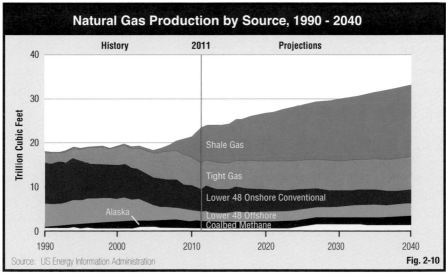

Natural Gas Production by Source, 1990 - 2040

History   2011   Projections

Shale Gas

Tight Gas

Lower 48 Onshore Conventional

Alaska   Lower 48 Offshore
          Coalbed Methane

1990   2000   2010   2020   2030   2040

Source: US Energy Information Administration

Fig. 2-10

the oil side of the tight sands equation, but the terminology gets confusing here, too. It typifies non-shale plays with low-permeability tight sandstone or limestone or carbonate reservoirs, sometimes interbedded with shales. Examples are the Niobrara oil play in Colorado and Wyoming, hybrid plays in the Anadarko Basin of Oklahoma and

Texas, and the Mississippi Lime chat (chert, dolomite, and limestone) play in Oklahoma and Kansas. These are important plays, too, but harder to lump into a standard category. Ironically, as some observers have seen the errors of their ways about equating tight oil sands plays with shale plays, they are adopting a more generic term to refer to all low-permeability oil plays as tight oil plays, lumping together shale and tight sands oil plays. As you'll see in the next chapter, such labels will become less important.

The 500-pound gorilla in any consideration of America's oil and gas picture, of course, is shale (Fig. 2-9). As important as tight gas and CBM have been to the US gas supply picture over the past few decades, they are dwarfed by the gargantuan scope of shale gas in any forecast (Fig. 2-10).

A breakdown of undeveloped, technically recoverable US shale oil and gas resources provided by Intek for EIA is shown in Table 2-1, published in 2011, but it includes my adjustment to the number for the Bakken Shale (doubled since then by the US Geological Survey with the addition of the equally huge underlying Three Forks/Sanish Shale formation). More generous estimates using a different methodology have been provided since then, notably by Advanced Resources International, a top consultancy for DOE, that puts the US shale gas and shale oil technically recoverable resources at 1,161 Tcf and 48 billion barrels, respectively.[40]

What Its Hydrocarbon Bounty
Means for America

Frankly, I could list a dozen different estimates for the potential oil and gas resources to be found in the US shale plays. It's a truism in this business that if you ask two petroleum geologists about resource estimates for a play, you'll get four answers (to be fair, you can also substitute reservoir engineers and reserves recovery rate in the same equation.)

---

40 http://www.eia.gov/conference/2013/pdf/presentations/kuuskraa.pdf

Estimated Remaining Undeveloped Technically Recoverable Shale Gas, Shale Oil Resources*	Trillion cubic feet	Billion barrels
Marcellus	410	--
Antrim	20	--
Devonian Low Thermal Maturity	14	--
New Albany	11	--
Greater Siltstone	8	--
Big Sandy	7	--
Cincinnati Arch	1	--
Haynesville	75	--
Eagle Ford	21	3
Floyd-Neal & Conasauga	4	--
Fayetteville	32	--
Woodford	22	--
Cana Woodford	6	--
Barnett	43	--
Barnett-Woodford	32	--
Avalon & Bone Springs	--	2
Mancos	21	--
Lewis	12	--
Williston-Shallow Niobraran	7	--
Hilliard-Baxter-Mancos	4	--
Bakken/Three Forks	--	7.3
Monterey/Santos	--	15
Total	750	27.3
		Table 2-1

*As of 1-1-2009, except for Bakken/Three Forks;
USGS revised estimate in 2013. Discovered shale resources.

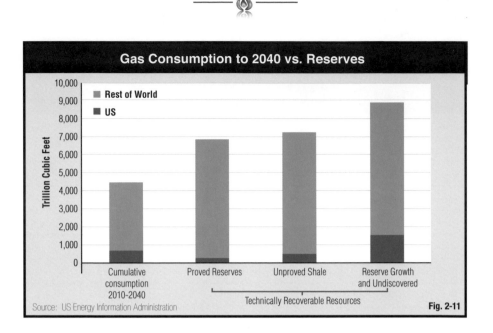

Gas Consumption to 2040 vs. Reserves

Source: US Energy Information Administration

Fig. 2-11

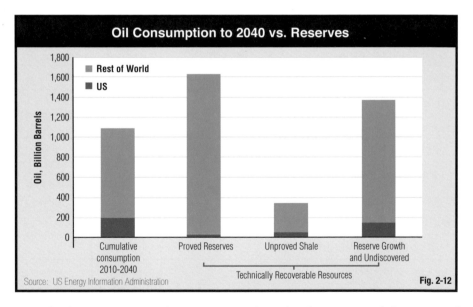

Oil Consumption to 2040 vs. Reserves

Source: US Energy Information Administration

Fig. 2-12

What's important is the context used in the discussion of the scope of these resources. Table 2-2 provides some of that context by illustrating how huge a boost cracking the shale "code" has been to America's oil and gas supply outlook. Both oil and gas get a comparable lift with the addition of shale resources to the US supply basket. But, relative to the US situation, shale oil is less important in a global context while shale gas is much more important. Of course, the vast conventional oil reserves of the Middle East, the oil sands of Canada, and extra-heavy oil of Venezuela have a lot to do with that differentiation.

Now check out Figs. 2-11 and 2-12, looking at cumulative consumption of oil projected from 2010 to 2040 for the US versus the world. Figure 2-12 suggests America will need every bit of its proved conventional oil reserves, potential oil from unproved shale/tight oil plays, and total undiscovered oil and incremental oil reserve growth (conventional and unconventional oil alike) to meet its expected oil consumption for the foreseeable future.

Given the uncertainty over oil prices and success rates, that points to a likelihood that our nation will continue to import oil. But a continuation of the unconventional oil growth and increased access to undiscovered resources could keep the level of oil imports inconsequential. There is no corresponding urgency for the rest of the world as a whole to tap shale oil, as proved conventional reserves outside the US alone can accommodate projected oil consumption. (Obviously, individual nations have their own agenda, irrespective of how much total conventional oil is out there.) And yet the US can rely more on its own resources to curb its voracious appetite for oil from volatile regions of the world, such as Nigeria, Venezuela, and the Middle East.

It's a different story on the natural gas side. Shale/tight gas potential plus proved reserves have us covered insofar as projected gas consumption goes, not to mention a big potential upside from reserve growth and undis-covered gas resources. While the reliance on proved gas reserves outside the US is not as heavy as it is for oil, it's important to note that more than two-thirds of those proven conventional gas reserves are concentrated in Russia, the Middle East, and North Africa—all sources prone to political manipulation of gas supplies (Russia) and/or violent upheavals.[41]

That sets the stage for the US to become a major exporter of its shale gas to other countries—particularly those seeking to curb their reliance on imports from volatile regions.

---

41 http://en.wikipedia.org/wiki/List_of_countries_by_natural_gas_proven_reserves

In a very real sense, then, the fracking revolution isn't just benefiting America directly by upending its energy outlook in a very positive way; it also may do a lot to ease energy security concerns for much of the rest of the world, reducing the potential of belligerent parties to use energy as a political weapon.

Technically Recoverable Shale Oil and Gas Resources in the Context of Total World Resources		
	Crude Oil (billion barrels)	Wet Natural Gas (trillion cubic feet)
United States		
Shale / Tight Oil and Shale Gas	58	665
Non-shale	164	1,766
Total	223	2,431
Increase in total resources due to inclusion of shale oil and shale gas	35%	38%
Shale as a percent of total	26%	27%
Total World		
Shale / Tight Oil and Shale Gas	345	7,299
Non-shale	3,012	15,583
Total	3,357	22,882
Increase in total resources due to inclusion of shale oil and shale gas	11%	47%
Shale as a percent of total	10%	32%
Source: US Energy Information Administration		Table 2-2

Bakken Shale frac job under way for Breitling Energy.

Chapter 3
Fracking: What It Is—and Isn't

Fracking is NOT...

Let's deal with a sore point about the fracking debate right away: what fracking isn't.

First, it is NOT a drilling process. Folks in the oil and gas industry tear their hair out every day because the mainstream media are, frankly, too lazy or technology-averse to bother to get this right. Not a day goes by where I don't come across a television, radio, newspaper, or magazine report covering the topic where it's referred to as a drilling process. Even some self-styled industry analysts who focus on the oil and gas sector (i.e., bloggers who are Wall Street analyst wannabes) fall into this trap.

This isn't splitting hairs just for the sake of argument. You really have to wonder how good of a job our industry is doing to educate the public on the most important, even revolutionary, oil and gas trends in a generation if no one outside of our industry understands this very basic distinction about a process that is fundamental to the current oil and gas boom.

Hydraulic fracturing (and the shortened term for the process is at issue too, per the next sidebar) does not take place until after drilling has concluded. Notice that I used the term "concluded" instead of "completed." There's a good reason for that, and it's another term folks outside the industry get wrong. A completion is a separate process that typically isn't even part of the drilling process. And, put simply, fracking is just another type of well completion process—a very expensive and complicated kind of completion, but a completion nevertheless.

Once a well has been drilled, the next step is to test and assess the target formation to determine if the well is to be completed in order to put it on production—or plugged and abandoned. The drilling crew installs and cements into place production casing in order to prepare the well for

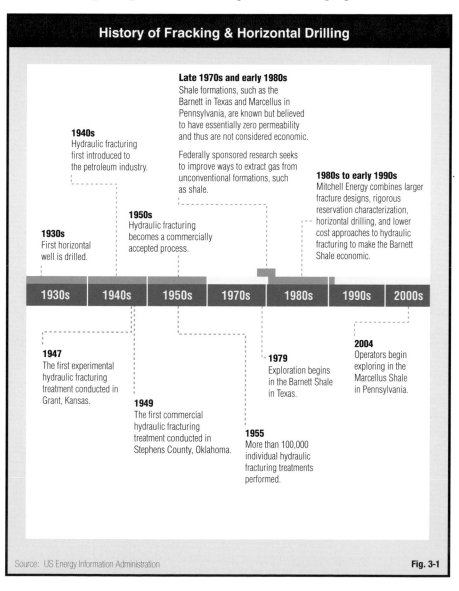

Source: US Energy Information Administration

Fig. 3-1

completion. The drilling rig then moves off the well location, and a service rig—a smaller rig purpose-built for completing, stimulating, or reworking wells—is brought in to undertake the completion. The service rig typically perforates the production casing and then runs production tubing, and the well is brought on line and begins producing, or it is sometimes capped temporarily and completed later.

To Frac or Not to Frack: How Spelling Sets an Agenda

Here I'm offering a caveat, mea culpa, challenge—whatever your preference is—to my colleagues in the oil and gas industry. Until hydraulic fracturing became a controversial topic, virtually no one in the industry spelled the short form of the term with a "k." You might have seen "frac," "fracing," "fraccing," or "frac'ing," even the rare "fract," but when hydraulic fracturing became a target of environmental activists, they adopted "frack." I suspect that many of them, having been blissfully unaware of the practice until opposing it became the cause du jour, immediately related it to the popular term "frak," a PG-rated substitute for a common profanity that was used on a popular television show of the time, *Battlestar Galactica*. Of course, this made for a colorful double entendre for protest signs that conjured up unwelcome associations with what the oil and gas industry does to Mother Earth. I've noticed folks in the industry make disparaging comments about this added "k" and what it signified, and they seem to have circled the wagons in opposition to this spelling. Well, this whole argument is so esoteric and trivial that it is lost on the public at large. In all my written communications on this topic, I've used "frack" and "fracking" mainly for the sake of convenience so that it doesn't distract us from the real issues. In fact, I would argue for industry to embrace the "k." Maybe we could work up our own protest signs and picket the Sierra Club headquarters, chanting, "Don't frack with America's economy and energy security." Just a thought.

("Later" can mean days, weeks, months, or even years. There are a lot of backlogged completions in the big "dry" gas shale plays awaiting higher natural gas prices to make them economic.)

Additionally, fracking is NOT something new or uncommon. It's believed that the first frack job was in the 1940s—depending upon whom you talk to, either in the Hugoton giant gas field in Kansas in 1947 or near Duncan, Oklahoma (former world headquarters for Halliburton), in 1949 (Fig. 3-1).

Furthermore, it's been commonplace for decades. Worldwide, it's estimated that more than 2.5 million wells have been fracked, and the US accounted for about half of those. Today, about 35,000 wells are fracked each year in all types of wells. And its impact on industry? It's been estimated that 80% of production from unconventional sources such as shales would not be feasible without it.[42]

Finally, before we get too deep into the details of the hydraulic fracturing process itself, there is a critical component of unconventional oil and gas development that gets overlooked in the discussion but is every bit as important as fracking: horizontal drilling.

Horizontal Drilling: The Other Key

It's ironic that as often as fracking is mistakenly referred to as a drilling process, the actual drilling process typically coupled to fracking in unconventional oil and gas development is just as often overlooked by the uninformed.

That's not to say that fracking does not occur in vertical and directional wells, too. It does. But in the context of economically developing our vast unconventional resources, horizontal wells are crucial. Technically, horizontal wells are actually a subset of directional wells; once the angle of a nonvertical well is greater than 80° from vertical, or more loosely, when the lower part of the well parallels the targeted pay zone, it is deemed a horizontal well. The whole idea is to expose

---

42 http://fracfocus.org/hydraulic-fracturing-how-it-works/history-hydraulic-fracturing

much more of the reservoir to the wellbore (Fig. 3-2). A typical vertical well in a shale play may be exposed to 50 feet of a formation versus a horizontal well's lateral extension, which may be exposed to thousands of feet within a formation that may be 50–300 feet thick. While most horizontal laterals are 2,000–6,000 feet long, some have been drilled in the Bakken Shale out to more than 10,000 feet.

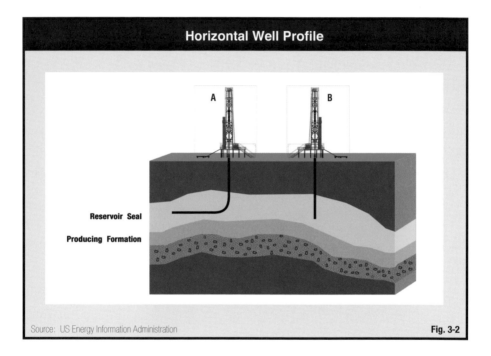

Horizontal Well Profile

Reservoir Seal

Producing Formation

Source: US Energy Information Administration

Fig. 3-2

Even before advances in fracking technology—coupled with improvements in horizontal drilling efficiencies—launched the unconventional resource boom, horizontal wells were delivering order-of-magnitude increases in oil production.

The first recorded truly horizontal well was drilled near Texon, Texas, in 1929. But there was little in the way of commercial application for horizontal drilling until the 1980s, when improvements in downhole drilling motors and downhole telemetry (getting current data about downhole conditions at the drill bit) greatly improved the speed and efficiency of drilling. Early adopters in the US included BP, working in Prudhoe Bay field, and many operating companies in the Austin Chalk formation of South Texas. About nine out of ten horizontal wells drilled in the world in the early 1990s were drilled in the Austin Chalk with unprecedentedly prolific results. Still, horizontal drilling accounted for no more than about 6% of all wells; today, that number is closer to 70%.[43]

And, amid all the controversy over the environmental impacts of fracking, the environmental benefits of horizontal drilling are often over-looked as well—chief among them the greatly reduced amount of surface disturbance. Full development of a one-square-mile section of a lease could entail as many as sixteen vertical wells, each situated on a sepa-rate concrete well pad. But drilling six to eight horizontal wells from a single well pad could access even more of the same reservoir volume. This is especially true of the shales, as their low permeability requires vertical wells to be placed even more closely than usual. In some plays, horizontal wells may disturb only one-tenth the surface area as vertical wells.

By requiring fewer locations for wells and thus disturbing less surface area, horizontal drilling also can significantly reduce the total number of well pads, access roads, pipelines, and surface production facilities. This helps minimize the overall environmental impact of the drilling operation by minimizing fragmentation of the habitat—a critical issue for wildlife concerns—as well as reducing dust, haze, noise, and road wear from truck traffic.

Horizontal drilling also allows development of resources underlying an area where a surface location isn't feasible—whether due to environ-mental sensitivity or practical reasons, such as being beneath existing structures—from a substantial distance away. A perfect example of this is the development of the Barnett Shale near Dallas-Fort Worth International Airport.

Some operating companies are pushing the envelope on horizontal wells with laterals that extend across multiple sections, or units. Often

---

43 http://gis.bakerhughesdirect.com/RigCounts/default2.aspx

multiple laterals, or multilaterals, are drilled horizontally from the same original vertical wellbore, resulting in "stacked laterals."[44]

Horizontal wells also must have the same precautions as do vertical wells to protect groundwater—the chief focal point of environmental concern in the debate over fracking. Figure 3-3 shows the multiple layers of protective steel casing and cement designed to protect freshwater aquifers, isolate the potable aquifers from saltwater zones, and to ensure that the producing zone is isolated from overlying formations.

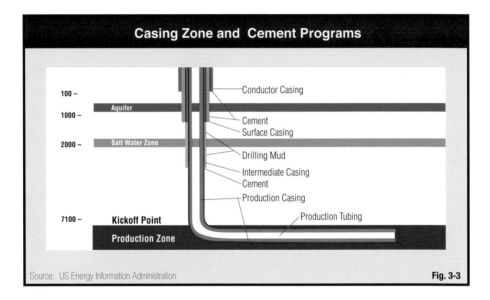

Fig. 3-3

Horizontal drilling efficiency has benefited tremendously from advances in drilling rig design. New-design rigs now have become marvels of drilling efficiency, with a wide array of automated features that are making a big difference in terms of improved safety and reduced rig downtime. Most noteworthy is the adoption of a top drive, which can replace the traditional rotary table and swivel in a conventional rotary rig (Fig. 3-4). The top drive is suspended from the top of the rig mast and rotates

44 http://www.google.com/url?sa=t&rct=j&q=&esrc=s&frm=1&source=web&cd=2&ved=0CC
  sQFjAB&url=http%3A%2F%2Fwww.aade.org%2Fapp%2Fdownload%2F7129847104%2F
  AADE%2B-Continental%2BRes.%2B-SCOOP%2B-Corey%2BRussell.pdf&ei=8O4cU8LY
  F9DiyAGymoCgCg&usg=AFQjCNHURG0gDEjvXVGHT8Trb9t_U_7Hjw

the shaft that holds the drill string. This powerful automated tool reduces the hands-on involvement of the rig crew and holds multiple joints of drill pipe at once instead of the usual one joint at a time.

This new breed of rig is quickly increasing its market share in the US as the unconventional oil and gas boom proceeds apace. Oil and gas companies working in some of the more operationally difficult plays, such as the Bakken Shale of North Dakota and the Eagle Ford Shale of South Texas, have been able to slash their total drilling days on a hole by 50% or more, thanks to these new rigs as well as their own progress up the play's learning curve.

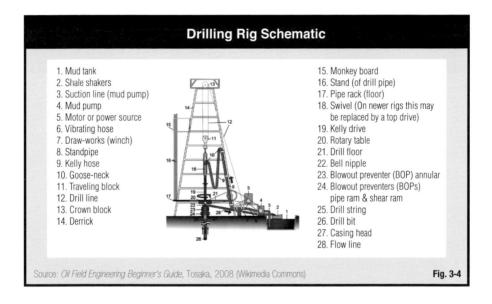

Drilling Rig Schematic

1. Mud tank
2. Shale shakers
3. Suction line (mud pump)
4. Mud pump
5. Motor or power source
6. Vibrating hose
7. Draw-works (winch)
8. Standpipe
9. Kelly hose
10. Goose-neck
11. Traveling block
12. Drill line
13. Crown block
14. Derrick
15. Monkey board
16. Stand (of drill pipe)
17. Pipe rack (floor)
18. Swivel (On newer rigs this may be replaced by a top drive)
19. Kelly drive
20. Rotary table
21. Drill floor
22. Bell nipple
23. Blowout preventer (BOP) annular
24. Blowout preventers (BOPs) pipe ram & shear ram
25. Drill string
26. Drill bit
27. Casing head
28. Flow line

Source: *Oil Field Engineering Beginner's Guide*, Tosaka, 2008 (Wikimedia Commons)   Fig. 3-4

Fracking: What It Is, How it Works

Hydraulic fracturing is essentially just another way to complete a well while stimulating the formation to produce more oil or gas. In effect, fracking creates additional permeability in a producing formation, enabling the oil or gas to flow more easily toward the wellbore. Usually that means overcoming existing barriers to the flow of hydrocarbons to the wellbore, whether they entail naturally low permeability common

to shale and tight sands formations or reduced permeability that results from drilling debris near the wellbore.

Fracking entails pumping a specially designed fluid into a formation at a preset pressure and rate to create cracks in the target formation (Fig. 3-5). In most of these unconventional formations, frack fluids are mainly water-based and mixed with additives that help the water transport sand (or tiny ceramic) particles into the fractures. Once the pumping process stops, these particles, or proppants, are used to prop open the fractures. Once the frack job has been initiated, more fluids are pumped into the well to continue expanding the fractures and deliver more proppant deeper into the formation; the pressure must be sustained downhole as the fractures keep getting longer.

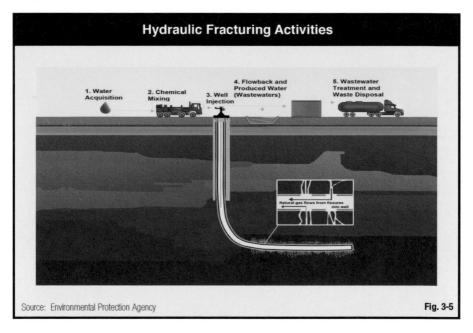

Frack Design

Designing frack jobs is a complex process that must take into consideration the characteristics of the targeted formation, such as the formation thickness and rock physics, fracture pressure, and intended fracture length. Refining these designs to optimize frack performance and boost hydrocarbon flow rates is a complicated, never-ending process. Not only do the frack designs vary a lot from play to play, they also vary from well to well.

Frack design incorporates a lot of high-tech wizardry. Chief among the high-tech tools employed are computer modeling to simulate the frack designs and microseismic fracture mapping and other downhole measurements to ascertain critical information such as the size and orientation of potential fractures. These tools help gather data that enables operating oil and gas companies to better manage resource development with improved placement of wells and by leveraging natural reservoir conditions to achieve the best fracture results in subsequent wells.

It's critical to optimize fracture patterns to make sure the fractures don't extend beyond the target formation, which under a best-case scenario could be just a waste of money and effort if a fracture extends into nonproductive rock, and under a worst-case scenario could damage the well and block off any oil or gas resources. Another concern regarding a wayward fracture is the inadvertent incursion of saltwater from an adjoining formation. Pumping extra water from the well hikes production costs and could squeeze the well's economics.

Fracking: The Process

The marriage of advanced horizontal drilling to hydraulic fracturing was the real turning point for the industry, notably in how oil and gas companies came to regard the practice of fracking itself.

Typically in the past, frack jobs in vertical wells (remember, there were a couple million of these jobs done long before anti-frack activists showed up for the television cameras) were single-stage affairs. Recall that a vertical well might perforate fifty feet or so of a shale formation, so even a successful vertical frack job might not yield economic flow rates as the fracture network couldn't access very much of the formation in a single well.

What really made things different was the shift to performing hydraulic fracturing in multiple stages that were essential for fracking long horizontal laterals. Typically, horizontal laterals in the shale plays extend from 1,000 feet to 6,000 feet and entail perhaps a dozen or more frack stages. There have been a number of frack jobs in some shale plays featuring as many as fifty to sixty frack stages along extended-reach horizontal laterals of 10,000–11,000 feet. That's drilling out horizontally about two miles after kicking off from a vertical well that may be 8,000–10,000 feet deep. Think about that for a moment. It's like starting a well at the inner base of Mount McKinley, North America's tallest peak, and drilling past the very summit!

Because the exposed wellbore in a typical horizontal shale well is so long, it is impossible to maintain enough pressure downhole to frack the lateral along its entire length in a single stage. So operators learned essentially to "partition" the lateral—isolating smaller portions of the lateral to accommodate a single stage each.

These multistage fracks are done in sequence, beginning with the farthest end of the wellbore and moving back uphole with each stage until the entire well has been fracked. Each frack stage entails a unique volume of frack fluids, with additives and concentrations of proppants specially designed for that specific stage.

Each of these stages is roughly equivalent to what you would find in a single vertical well frack job. So you can imagine how mind-blowing it must be to oil field service company veterans to encounter hydraulic fracturing that amounts to stimulating fifty or sixty wells all at once.

(Of course, just knowing that oil and gas companies are now producing huge volumes of oil and gas from source rock probably already blew their minds.)

Before the actual frack job begins, of course, the operator orders tests of the wellbore casing and cement and the fracking equipment to ensure that all can handle the enormous fracking pressures and high-horsepower pump flow rates. In all instances, the wellbore construction must comply with a host of state oil and gas regulations to make sure the well is safe and any ground or surface water isn't contaminated.

Prior to a frack stage, the near-wellbore area is cleaned up with an acidic solution, typically to unplug rock pores plugged by drilling mud or casing cement. Then a water-based frack fluid is mixed with a specially designed agent to reduce friction and thus expedite the flow and placement of the proppant. Then large pumper trucks pump a huge volume of water into the wellbore, along with the finely grained proppant. In subsequent stages, the volumes of proppant are gradually increased while the overall volume of the fluid slurry decreases. Subsequent stages typically contain a coarser proppant. After the final stage is completed, the well is flushed with enough fresh water to remove excess proppant.

Staging frack jobs this way enables the operator to tightly control the fracking process, making modifications to accommodate formation changes along the wellbore route.

Once the frack treatment is complete, the oil and gas are released through the new fractures into the horizontal lateral, the well pressure is released, and the well stream flows up the vertical portion of the well. The well is capped with a wellhead at the surface, the well stream is treated to remove the frack fluids and wastewater (a mix of naturally occurring fluids and elements in the fracked formation and frack flowback water), and the hydrocarbons are prepared for delivery into a pipeline gathering system—assuming there is a market for the oil or gas.

Frack jobs do consume a large volume of water—perhaps 2–4 million gallons with each job. That's equal to about three to six standard-size swimming pools. In areas stricken by drought, all water uses have to be viewed with concern. But let's put things in perspective. That's about how much water a golf course uses in a typical summer month. According to the US Geological Survey, water withdrawals by electric utilities account for about half of the hundreds of billions of gallons used in the US each day. Generally speaking, water use by all mining activities (a category that includes oil and gas drilling, completion, and production) is about 1%.[45]

But it can be a problem for the service company handling the frack job to secure such a large volume of water, especially in areas where water resources are already scarce. That's why oil and gas companies are moving increasingly to recycling wastewater from previous frack jobs.

---

45 http://pubs.usgs.gov/fs/2009/3098/

While recycling wastewater is a big, expensive job, dealing with it otherwise has its challenges, too. The options besides recycling entail treatment followed by surface discharge or disposal by underground injection.

Fracking and directional or horizontal drilling are just tools, greatly improved in recent years, that we use to recover the oil and gas that power our economy—and the oil and gas industry that has been one of the few engines of growth in the US economy since the Great Recession began. I don't think it's exaggerating to suggest that a fracking ban, as some have proposed, would utterly decimate an industry that accounts for about 10 million jobs in the US and represents about 8% of US GDP.

And yet there are some who seem to be willing to do just that by promoting their agenda with little regard for the actual facts and science. It frustrates me to wonder how they can't see that the potential results of their actions might be even worse for the environment, not to mention America's economic and energy security, than the practice they so ardently oppose.

The unsubstantiated fear-mongering and campaign of disinformation in service to that agenda warrants a clear-headed round of myth-busting in the next chapter.

Breitling's Buffalo Run Well undergoing a frac job
with water supply pit in foreground.

Chapter 4
Fracking: Busting the Myths

Energy Boom's Unexpected Parentage

The current American energy boom boasts a parentage that might surprise some of the folks who worry about how fracking might damage the environment and want the federal government to regulate the practice more tightly if not ban it altogether.

Yes, today's energy revolution was the brainchild of a staunch environmentalist whose work was, in a manner of speaking, "midwifed" by the US government. Okay, I'm stretching a point a bit, but bear with me.

The catalyst for the key innovations in hydraulic fracturing that launched tens of thousands of unconventional oil and gas wells was the stubbornness of one man: George P. Mitchell. The son of a Greek immigrant and eventual Texas A&M petroleum engineer made as much of a case for supporting sustainable development and environmental values as he did for promoting the oil and gas industry. Mitchell, who died a billionaire in 2013 at the age of ninety-four, built Mitchell Energy & Development Corporation, a successful and mostly natural gas-focused independent oil and gas company that was also a world-class real estate developer.

George Mitchell believed passionately in the potential of the Barnett Shale, a vast shale formation underlying much of the Dallas-Fort Worth metropolitan area in the Fort Worth Basin in North Central Texas. Bucking the naysaying of his peers in the industry as well as his own board, Mitchell spent fifteen years and $6 million to "crack the code" of the Barnett Shale through improvements in hydraulic fracturing. Constantly being told he was wasting his time and money and even risking having his company taken away from him, Mitchell stubbornly persisted with his fracking experiments, begun in the early 1980s. Mitchell's tests of various frack fluids had resulted in a breakthrough in 1997, drilling was stepped up, more operators joined the play, and within five years Barnett Shale gas production had grown tenfold.

After Devon Energy bought Mitchell Energy & Development for $3.5 billion in 2001, its own aggressive pursuit of improvements in horizontal drilling in tandem with the newly enhanced frack technology yielded still more step-changes in drilling activity and production growth. By the mid-2000s, the Barnett Shale was accounting for about 7% of total US gas production and has still maintained output levels near 5 billion cubic feet per day. It is expected to yield ultimately about 45 trillion cubic feet of gas (Tcf).[46]

Mitchell had some help along the way, and it might irk some of my colleagues in the industry to be reminded of the role the US government played in the R&D that helped further the great shale fracking breakthrough.

Federal efforts to research ways to boost America's natural gas production by targeting unconventional gas resources go back to the 1970s. Industry was well aware of these huge resources but deemed them economically nonviable with current technology. Consequently, Congress passed legislation establishing the Unconventional Gas Research Program in 1976. Over the fifteen-year life span of the program, federal funds in support of basic R&D into unconventional gas—notably the Eastern Devonian shales (shales developed 350 million years ago during the Devonian period)— totaled $92 million. US Department of Energy-supported research included studies in these formations' geochemistry, geology, resource assessments, and the like. Especially critical was the development of microseismic monitoring of frack treatments that would not have been possible without the

---

46 http://blogs.star-telegram.com/barnett_shale/barnett_shale_production/

DOE-funded research in that area.[47] Progress was also made in advancing horizontal drilling methods, geologic mapping, and downhole telemetry.

Ironically, it was the widely despised Windfall Profits Tax Act of 1980 that provided a price support for unconventional gas production via a product tax credit. While that credit was in effect from 1980 to 2002, production of unconventional gas in the US more than doubled (Fig. 4-1).[48]

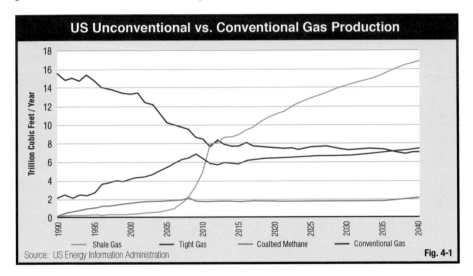

US Unconventional vs. Conventional Gas Production

Source: US Energy Information Administration

Fig. 4-1

How important was the federal role in "midwifing" the advances that proved critical to exploiting unconventional gas resources? According to Fred Julander, past president of the Colorado Oil & Gas Association, "The Department of Energy was there with research funding when no one else was interested, and today we are all reaping the benefits. Early DOE R&D in tight gas sands, gas shales, and coalbed methane helped to catalyze the development of technologies that we are applying today."[49]

Now there's been a bit of controversy over how important the role of the federal government was in catalyzing the development of unconventional gas—even to the extent that some have claimed Mitchell could not have succeeded without government support. And it's a bit of a misnomer to

---

47 http://americanenergyinnovation.org/staff-research/case-study-unconventional-gas-production-2013/
48 http://www.eia.gov/energy_in_brief/data/natgas_production_AEO2013ER.xlsx
49 National Energy Technology Laboratory. "Shale Gas: Applying Technology to Solve America's Energy Challenges." US Department of Energy. March 2011.

call Mitchell the "Father of Fracking" when in fact fracking was well established before the Barnett Shale breakthrough. Mitchell clearly benefited from the DOE-funded research, especially with regard to microseismic mapping. But it's pretty clear the breakthrough, a means to commercially develop gas from shale via fracking, was something no one else—not the government, even with the millions directed toward unconventional gas R&D—had accomplished.[50]

In 1997, when George Mitchell's brainchild was "delivered," US proved reserves of natural gas were estimated at only 164 Tcf. At the time, total US gas consumption was about 22 Tcf per year. That means total proved US reserves at that time would have lasted only about seven years. Today, we are consuming about 26 Tcf per year, but thanks to George Mitchell, we now have a technically recoverable gas resource that can last ninety-two years at current consumption levels (Fig. 4-2)[51]. Of course, as I explained earlier, comparing proved reserves to resources is comparing apples to oranges, but in this case, given the near certainty of technical producibility of the shale gas resources, I'm confident that a moderate, sustained increase in natural gas prices will see the lion's share of those reserves get booked as proven.

The Fracking Environmentalist

Here's the thing about George Mitchell: He was much more than an oil and gas man or even the misnamed "Father of Fracking." He was a pioneering environmentalist as well. Case in point: Mitchell's ambitious, groundbreaking—and wildly successful—experiment in smart, environmentally friendly urban development, The Woodlands, Texas, north of Houston. Mitchell championed the concept of developing a master planned urban community that minimized land use in an environmentally conscious, sustainable way. The Woodlands showed how to create a smarter urban community by conserving open spaces and parkland, providing a range of options in living and working environments, accommodating multiple means of transportation modes, and promoting collaborative decision-making between public and private sectors.

---

50 http://cgmf.org/blog-entry/75/How-much-did-the-Feds-really-help-George-Mitchell.html
51 http://www.eia.gov/tools/faqs/faq.cfm?id=58&t=8

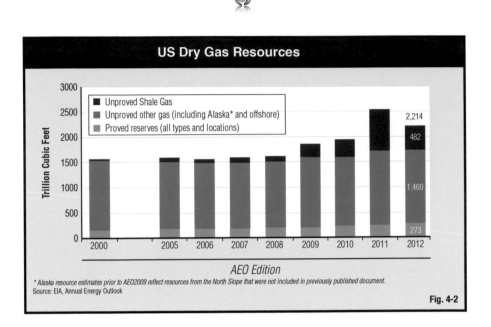

US Dry Gas Resources

	Trillion Cubic Feet
■	Unproved Shale Gas
■	Unproved other gas (including Alaska* and offshore)
■	Proved reserves (all types and locations)

2,214
482
1,460
273

2000 2005 2006 2007 2008 2009 2010 2011 2012

AEO Edition

* Alaska resource estimates prior to AEO2009 reflect resources from the North Slope that were not included in previously published document.
Source: EIA, Annual Energy Outlook

Fig. 4-2

As Harry H. Frampton III, former chairman of the Urban Land Institute and managing partner of East West Partners, put it, "George figured it out long before the rest of us, but we cannot keep using more land than is necessary for development and expect to preserve land at the same time."[52]

Mitchell's environmental plaudits don't end there. He helped create a forty-five-acre habitat for the endangered whooping crane near the Aransas National Wildlife Refuge in Texas. He's provided six-figure prizes for entrepreneurs who achieve significant progress in conservation of resources. He lobbied for the phase-out of single-hulled tankers due to their increased risk of catastrophic oil spills.

Mitchell had no tolerance for his colleagues who didn't respect environmental values. In one of his last interviews before he died in 2013 he said that "wild" operators need to be well regulated. "If they do something wrong and dangerous, they should punish them," he told a reporter. In the mid-1970s, in the midst of an energy crisis, he founded the Woodlands Conference Series to encourage the study of sustainability, which he later accurately predicted, "will be one of our most serious problems. Energy is

52 http://www.chron.com/opinion/outlook/article/Taylor-The-Woodlands-set-land-use-standard-4705296.php

part of it. The global climate is part of it. We need to do solar and wind, but they're not enough. Now we've got 6.5 billion people. In 2050, we'll have another 3 billion. If you can't make it work now, you've got more wars, more poverty, and everything you can think of."[53]

But Mitchell was a bit of a paradox, as would befit an environmental activist whose business is fossil fuels.

Longtime Houston energy writer Loren Steffy, in a commemorative essay about Mitchell for *Forbes* magazine, noted that:

Mitchell never saw any irony in his dual role as the father figure of both fracking and sustainable development. He often criticized the environmental movement for not going far enough, for stopping short of supporting sustainable development. At the same time, his company, Mitchell Energy, was not a "green" business. While he followed environmental regulations—and in some cases took initiatives to set aside some of his company's oil and gas leases to protect endangered species—Mitchell Energy didn't reflect his own beliefs on sustainability.

It's with fracking, though, that the Mitchell Paradox finds unity. By unleashing the huge domestic resource of shale gas, Mitchell provided a path to reducing the growth of carbon emissions while enhancing the country's energy security. Mitchell understood the importance that continued energy development played in our country's future, yet he also recognized the tremendous responsibility we have to protect the environment as the human population grows ... Fracking offers a greener alternative, one that benefits society, the environment and the oil business. Mitchell's legacy is creating that bridge.[54]

I would not be so presumptuous as to claim myself an heir to the George Mitchell legacy of fracking intertwined with sustainability, but I certainly am committed to emulating him in the pursuit of both toward a common

---

53 http://www.chron.com/business/energy/article/George-Mitchell-still-pushes-energy-conservation-1678366.php
54 http://www.forbes.com/sites/lorensteffy/2013/07/29/remembering-george-mitchell-and-his-paradox/

objective—a future that embraces both energy security and sustainability. And I sincerely believe that, contrary to claims that it is an environmental menace, fracking will help keep us on the path to that future.

But first we have to bust some myths about fracking.

Hollywood and the Misinformation Machine

Every country has a stake in how the energy boom unfolds, and each knows it. About a third of my time is spent traveling around the world researching and speaking about the fracking revolution. In great cities such as Beijing, Singapore, Istanbul, Berlin, Amsterdam, and Rio de Janeiro and in emerging economies such as Thailand and Myanmar, audiences are eager to learn everything they can about fracking, how to do it, and what it means for their culture, economy, and environment.

Reducing high energy costs, securing a reliable supply of home-grown fuel, and dealing with pollution are life-and-death issues that all the wind and solar farms in the world won't be able to solve for a long time.

We now know that we live on top of a supply of hydrocarbons that by some estimates could sustain us for centuries, and in exploiting it we are among the most vigilant protectors of the environment. The Wild West of sacrificing environmental quality for economic progress isn't in America anymore. It's in exploding economies such as China and India, where oversight and regulation take a back seat to finding enough power and raw materials to keep billions of people fed, housed, and employed.

In the US, every aspect of oil and gas production is closely watched, hotly debated, and highly regulated. We have a free and open press that is capable of sorting fact from fiction. Yet it's here at home, in the cradle of freedom and opportunity, that some of the most ridiculous ideas, myths, and misguided fears about the technology have been born and propagated. Unfortunately, as I found in my travels abroad, such misinformation has proven to be a successful export.

I was on my way to a conference in Calgary, Alberta, Canada, when the cab driver who picked me up at the airport asked me the purpose of my visit. This is roughly how the conversation went:

"It's an oil and gas conference."

I saw in the rearview mirror that his eyes narrowed.

"Do you know what fracking is?" he asked.

"Sure, I do."

"Do you do fracking?"

"Yes, I do."

"Where?"

"Texas, and all over the country."

"Well, I have a really big bone to pick with you."

Oh, great, I thought. Here we go again. In my field, a lot of people want to pick bones with me. This was not long after the big earthquake and tsunami in Japan that killed nearly 16,000 people and caused the Fukushima nuclear power plant catastrophe.

"You know," the driver began, "I cannot believe how you could continue to do this fracking. You guys down there in Texas caused that earthquake in Japan and killed all those people."

He continued his tirade all the way to my hotel. Nothing I said could change his mind.

Fracking practices do occasionally cause microseismic events immediately around active wells, and fracking has been linked to—but not proven conclusively to be the cause of—a few minor tremors in Oklahoma, England, and British Columbia. But it's just delusional to think that some pinpricks in the ground in North America could cause a massive shift in tectonic plates half a world away.

Even in the few instances where linkage to earthquakes was firmly established, it was the drilling of injection and disposal wells mandated by regulatory agencies to dispose of frack flowback water that was linked to the tremors—not the fracking process itself.

A couple of years ago I showed up at a fracking conference where I was to give a speech in Warsaw, Poland, and found that a group of protesters had chained themselves together in a circle on the stage. There had been no wells fracked in Poland yet, so no one there had any direct experience of or specific complaints about the practice. While the conference orga-

nizers cajoled the protesters to leave, I talked to some of them. They were respectful, thoughtful people, but convinced by what they'd read on the Internet that fracking would destroy their land, "like in America."

I heard a similar argument when I spoke in South Africa. Protesters there told me, "We know that you guys have destroyed pretty much every place you've drilled in America and forced people to sign nondisclosure agreements to hide what's going on."

How did the greatest energy discovery since 1859, when oil first gushed out of the ground in Titusville, Pennsylvania, become so cartoonishly controversial and misunderstood?

It helps to put it in context, and that context is the highly unfortunate confluence of two events: the BP Deepwater Horizon drilling rig disaster and resulting catastrophic oil spill in the Gulf of Mexico, and the release of the anti-fracking documentary film *Gasland*.

Not only was Deepwater Horizon the second-worst oil spill in history (after the deliberate sabotage of oil wells in Kuwait during the Gulf War in 1991), it happened in our backyard, not in some distant country with no press freedom.

Thanks to modern communications, tens of millions of us got to watch the catastrophe unfold in real time. Satellite imagery allowed us to see exactly how widespread the damage was, and undersea cameras gave us a depressing, front row view of the billowing clouds of brown gunk spewing from the well. We also had a real-time view of attempts to shut in the runaway well, many of us sitting on the edge of our seats hoping each time it would work.

Until Deepwater Horizon, resentment of the oil industry had simmered mostly over prices and profits. The BP disaster graphically exposed the risks of our dependence on oil. BP's handling of it—the CEO suggesting at one point that the impact wouldn't be all that bad and at another point complaining that he'd like his life back—proved to be salt in the wound. Even if BP had reacted more quickly and embraced its responsibilities earlier, how could anyone finesse images of dying and dead pelicans mired in thick black goop, or interviews with anguished fishermen and hotel owners who'd been put out of business?

The Deepwater Horizon explosion and spill began April 20, 2010. Sixty days later, with the BP well still out of control, still gushing millions of gallons of fudge-colored crude and still front-page news, a low-budget documentary claiming to be an exposé of natural gas fracking in Pennsylvania had its television premiere on HBO.

Gasland had been a minor curiosity at the Sundance Film Festival earlier that year. By the time it aired in June, the oil business faced a broad and vitriolic hate-fest fueled by intense media coverage. Director/producer/star Josh Fox's flawed but provocative film—most famously its video depiction of flames shooting out of water faucets—captured the imagination of a public that was now primed to buy into sweeping claims of water pollution, breast cancer clusters, earthquakes, and the like.

In the blink of an eye, Josh Fox became the darling of the environmental movement, picking up celebrity boosters such as actress Debra Winger and John Lennon's widow, Yoko Ono. With Hollywood aboard and the mainstream media saturating our news with images of flaming tap water, the film won more attention and a patina of credibility. The ripples spread far and wide, sweeping up every harebrained theory along the way, including responsibility for the earthquake in Japan.

That's how cab drivers in Canada, farmers in Poland, and environmentalists in South Africa got the idea that we're turning America into a toxic wasteland.

There is irony to spare in this story. In stirring up such a huge backlash against all fracking by anybody anywhere, *Gasland* made it harder for journalists and the public to differentiate between companies that were operating responsibly and those that were not.

I am not saying that every claim in *Gasland* was wrong. There had, indeed, been problems and legitimate concerns. In Pennsylvania, some less-experienced operators installed inadequate well casings, posing a risk to groundwater supplies; others were careless with the handling of the millions of gallons of frack flowback water. Country roads and bridges designed to handle light local traffic groaned and crumbled under the constant pounding of loaded tanker trucks and heavy construction equipment, which during dry spells stirred up clouds of dust.

But as the *New York Times* wrote in its review of *Gasland*, the film-maker relied heavily on "vivid images—bright red Halliburton trucks, beeping but unidentified scientific instruments—over the more mundane crossing the t's and dotting the i's of investigative journalism. It's maddening to see how easy he makes it for the film's critics to attack him, and how difficult for sympathetic but objective viewers to wholly embrace him."[55]

The most effective response to *Gasland* came not from the industry but in a competing documentary film released in January 2013. *FrackNation*, by Irish journalist and film documentarian Phelim McAleer and his wife, Ann McElhinney, exposed a lot of holes, misstatements, and exaggerations in Fox's film. In its review, the *New York Times* wrote, "*FrackNation* is no tossed-off, pro-business pamphlet. Methodically researched and assembled, Mr. McAleer knows his way around the Freedom of Information Act and has done his legwork. More than anything, *FrackNation* underscores the sheer complexity of a process that offers a financial lifeline to struggling farmers."[56]

While Fox was embraced by the environmental pressure groups and the media, McAleer became the darling of the right, hailed and celebrated by Tea Partiers and like-minded conservative groups, but largely ignored by the media. His film was factually credible, but he'd already been tagged as a climate-change denier because of an earlier film he had made challenging some of the facts in Al Gore's popular documentary film *An Inconvenient Truth*.

About Those Flaming Faucets...

Ground Zero for the *Gasland*-spawned controversy over fracking and flaming faucets was Dimock, Pennsylvania, where a well operator—Cabot Oil & Gas—was blamed for methane gas contaminating local wells and where Josh Fox filmed the memorable footage of faucets spewing flames.

---

55 "The Costs of Natural Gas, Including Flaming Water," by Mike Hale, *New York Times*, June 20, 2010

56 "A Flip Side to the Attack on Fracking," by Jeannette Catsoulis, *New York Times*, January 10, 2013

The controversy prompted the federal Environmental Protection Agency to spend a few million dollars testing drinking water in the area. Hard as it tried, the agency's search for gas contamination that could be traced to fracking came up empty-handed. That there would be methane in water where there is an underground formation spanning thousands of square miles and holding massive volumes of the gas—the Marcellus Shale that covers much of Pennsylvania and neighboring states—doesn't surprise geologists. The water does contain biogenic methane—naturally occurring gas that's dissolved in water and bubbles out when water is pumped to the surface.

It's well-established science that oil and gas are constantly seeping up out of the earth without any help from man or machine—far more than spills and releases due to human error. It's also well-established folklore: None other than Founding Father and first President of the United States George Washington and noted revolutionary Thomas Paine were said to have set fire to the Rocky Hill millpond in New Jersey to demonstrate the natural release of methane from the pond bed. And it's been a tourist attraction: Until 1959, children could splash under a flaming water fountain in Colfax, Louisiana, where artesian water wells contained so much naturally occurring methane that the water could be lit afire. Sound familiar?

Gasland's claim that fracking caused gas to enter the water supply in Dimock, Pennsylvania, was proven false, but that was never as widely or vividly reported as those burning faucets. It's no surprise that a flaming kitchen tap is what most people remember.

Lest you think that Fox attempted to remedy any of *Gasland*'s misleading or downright untruthful information and footage in the sequel, *Gasland 2*, I'd just like to note that scenes in the sequel showing a garden hose spewing fire were not just misleading but actually faked: the hose was hooked directly to a gas vent.

So, instead of taking the word of a filmmaker whose claims have been widely discredited, let's review a few of the more reputable attempts at finding out whether or not fracking causes methane seepage into drinking water.

The Canadian province of Alberta, where more than 100,000 gas wells were drilled in the mid-2000s, was concerned enough in 2006 that it started an extensive monitoring program of 126 water wells. About a third did have methane in the water, but the provincial government reported five years later that none contained methane gas that resulted from drilling.[57]

A Pennsylvania State University study[58] of water quality in the heavily drilled Marcellus Shale compared pre-drilling and post-drilling water chemistry and concluded, "When comparing dissolved methane concentrations in the 48 water wells that were sampled both before and after drilling (from Phase 1), the research found no statistically significant increases in methane levels after drilling and no significant correlation to distance from drilling."

A similar phenomenon occurs in the oceans. In 2003, the US National Academy of Sciences estimated in its "Oil in the Sea III" report that there are more than 600 natural oil seeps in the Gulf of Mexico alone that leak as much as 5 million barrels of oil per year. Globally, the Academy estimated the figure at 15 million barrels. Considering that the oceans are largely unexplored, it's safe to say the figure is probably a lot higher.[59]

Whereas oil seeps are easy to spot because crude has a strong odor and it sticks to everything, natural gas is odorless and invisible. No one knows how much natural gas is bubbling to the surface or escaping from natural land seeps. None of this gets the industry off the hook, but it helps put the issue in perspective.

According to Terry Engelder, a professor of geosciences at Pennsylvania State University and the man credited with first identifying the potential of the Marcellus Shale, experts believe the techniques used in the beginning of the shale boom "were just inadequate to the task."

---

57 http://environment.gov.ab.ca/info/posting.asp?assetid=8136&categoryid=5
58 http://www.rural.palegislature.us/documents/reports/Marcellus_and_drinking_
   water_2012.pdf
59 http://en.wikipedia.org/wiki/Petroleum_seep

Engelder has said the industry should sit down with opponents, explain the geological science, the risks, and the benefits. "I would do whatever it took to try and engage these people over a period of time," he told a reporter.[60]

With the *Gasland* controversy swirling around it, Cabot Oil did begin holding summer picnics in the Dimock area to answer residents' questions. The company said more than 8,000 people attended in 2012. But the damage to fracking's reputation had already been done.

Government's Role

What role has government played in resolving the fracking controversy? Not much and not well at the federal level is the inescapable answer.

Unfortunately, the federal Environmental Protection Agency (EPA), a quasi-cabinet-level agency in the best position to sort it all out, has fumbled the opportunity to lead. For example, the agency issued a draft report in 2011 based on testing of contaminated water wells near drilling and fracking operations in Texas, Wyoming, and Pennsylvania. The report said some pollution found in shallow water wells was probably the result of seepage from old waste pits nearby, but it also claimed that chemicals found in deep test wells were residue from frack fluids.

The report met a storm of criticism from the drilling industry as well as state regulators, challenging the EPA's interpretation of the data. The EPA beat a hasty retreat, announcing it would conduct a national study that wouldn't be done until 2016—basically kicking the can down the road. Fracking techniques have been in use for decades. Nine out of ten new wells drilled in the US are fracked. The EPA has had oversight of aspects of the oil and gas industry for years. It should have been out front on the issue, not chasing it.

Another agency that's playing catch-up is the Interior Department's Bureau of Land Management, which is writing new rules for fracking on public and tribal lands. Most drilling takes place on private lands, but when

---

60 *FrackNation*, Feature Documentary by Phelim McAleer and Ann McElhinney, released January 2013

the federal government sets a rule for itself, the states usually follow. So the agency has become a political flashpoint in the all-or-nothing style of debate that dominates the fracking controversy. An estimated 1 million public comments were submitted, and the responses came as expected, along the same old divisions—environmentalists want no fracking anywhere, ever, and the industry opposes what it sees as excessive regulation.

As a result, instead of clarity and objectivity, we ended up with just another skirmish in the civil war over energy, and it continues unabated.

What's in That Stuff?

The current fracking boom has been fueled in part by passage of the Energy Policy Act in 2005, which, in addition to providing tax incentives and loan guarantees for energy production, exempted fracking from the reporting requirements of the Safe Drinking Water Act as well as the Clean Air Act and Clean Water Act.

According to a 2010 report by the Society of Petroleum Engineers, fracking regulation was left to the states and "has been piecemeal. This has allowed the energy industry to keep the chemicals used in fracking fluids secret."[61]

There have been several attempts to roll back the secrecy exemption. The proposed Fracturing Responsibility and Awareness of Chemicals Act (FRAC) has been introduced in three successive congressional sessions (2009, 2011, and 2013), and each time the bill was buried in committee. According to the website GovTrack.us, the proposed legislation has about a 1-in-100 chance of passage anytime soon.

The argument for secrecy is that oil field service companies are experimenting with different formulas for more effective frack jobs. Any company that's got a recipe that works doesn't want its competitors to know how they did it, or how they did it cheaper or faster than the next guy.

Keeping trade secrets is legitimate in any business, but keeping secret the chemical composition of frack fluids was a gift on a silver platter for

---

61 "The Shale Gas Boom: A Fracking Primer," edited by Dr. Mir F. Ali, p. 11; mirfali.files. wordpress.com/2013/08/fracking_chris.doc

people looking for reasons to target the oil and gas industry on environmental issues. For starters, the exemption became derisively known as the "Halliburton loophole" because Vice President Dick Cheney pushed for its passage. Cheney was a former CEO of Halliburton, one of the world's largest oilfield service companies. It was signed into law by President George W. Bush, who was also an oilman.

The Halliburton loophole secrecy exemption plays well for those with a conspiracy bent who are also eager to catch the oil and gas industry at a practice that sounds unsavory in terms of environmental impact. Opponents of fracking have resorted to some pretty extreme language when they try to describe the process. Some activists have conjured up an image of well frackers akin to mad scientists stirring a witch's brew in the bowels of the earth, then pumping it back to surface, sometimes in our backyards. Perhaps most egregious are their references to a "toxic cocktail" or "toxic soup" of chemicals "blasted" downhole to threaten nearby drinking freshwater sources.

Any scientist worth his or her salt will tell you that it's the dose that makes the poison, even in today's overly litigious, zero-risk climate. Even air or water can kill if you overdose on these life essentials. This kind of hyperbole has been around for decades. For example, consider

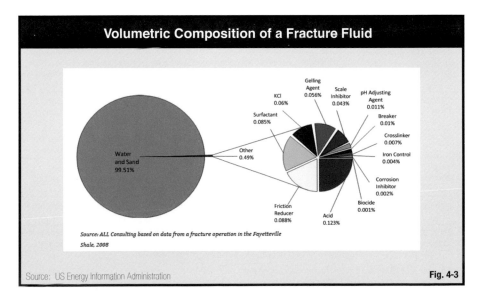

the great saccharin scare of the 1970s, in which government warning labels had to be attached to the artificial sweetener because it was linked to bladder cancer in rats.

FRACTURING FLUID ADDITIVES, MAIN COMPOUNDS, AND COMMON USES			
Additive Type	Main Compound(s)	Purpose	Common Use of Main Compound
Diluted Acid (15%)	Hydrochloric acid or muriatic acid	Help dissolve minerals and initiate cracks in the rock	Swimming pool chemical and cleaner
Biocide	Glutaraldehyde	Eliminates bacteria in the water that produce corrosive byproducts	Disinfectant; sterilize medical and dental equipment
Breaker	Ammonium persulfate	Allows a delayed break down of the gel polymer chains	Bleaching agent in detergent and hair cosmetics, manufacture of household plastics
Corrosion Inhibitor	N,n-dimethyl formamide	Prevents the corrosion of the pipe	Used in pharmaceuticals, acrylic fibers, plastics
Crosslinker	Borate salts	Maintains fluid viscosity as temperature increases	Laundry detergents, hand soaps, and cosmetics
Friction Reducer	Polyacrylamide	Minimizes friction between the fluid and the pipe	Water treatment, soil conditioner
	Mineral oil		Make-up remover, laxatives, and candy
Gel	Guar gum or hydroxyethyl cellulose	Thickens the water in order to suspend the sand	Cosmetics, toothpaste, sauces, baked goods, ice cream
Iron Control	Citric acid	Prevents precipitation of metal oxides	Food additive, flavoring in food and beverages; Lemon Juice ~7% Citric Acid
KCl	Potassium chloride	Creates a brine carrier fluid	Low sodium table salt substitute
Oxygen Scavenger	Ammonium bisulfite	Removes oxygen from the water to protect the pipe from corrosion	Cosmetics, food and beverage processing, water treatment
pH Adjusting Agent	Sodium or potassium carbonate	Maintains the effectiveness of other components, such as crosslinkers	Washing soda, detergents, soap, water softener, glass and ceramics
Proppant	Silica, quartz sand	Allows the fractures to remain open so the gas can escape	Drinking water filtration, play sand, concrete, brick mortar
Scale Inhibitor	Ethylene glycol	Prevents scale deposits in the pipe	Automotive antifreeze, household cleansers, and de-icing agent
Surfactant	Isopropanol	Used to increase the viscosity of the fracture fluid	Glass cleaner, antiperspirant, and hair color

Note: The specific compounds used in a given fracturing operation will vary depending on company preference, source water quality and site-specific characteristics of the target formation. The compounds shown above are representative of the major compounds used in hydraulic fracturing of gas shales.

Source: DOE National Energy Technology Laboratory   Table 4- 1

Well, as it turned out, the tumor-growing rats had been force-fed huge volumes of the stuff, equal to a human drinking bathtubs full of diet soda each day. Which, of course, by itself would kill you on the spot.

So what's really in the frack fluid slurry? This "toxic soup" that gets "blasted" downhole comprises the following: 90% fresh water, 9.5% sand, and 0.5% chemical additives.

Okay, so what are the chemical additives? Most of the stuff in there is in trace amounts and is mainly there to reduce friction and prevent corrosion, as Figure 4-3 shows.

But some of that stuff—acid, surfactant, biocide—might sound a little scary to the layperson. So check out Table 4-1 to see what each of these additives is used for in a frack job and in everyday life.

Sure, you wouldn't want to drink antifreeze (ethylene glycol) or swimming pool cleaner (muriatic acid), but there are only tiny, even trace, amounts of these substances injected into a rock formation more than a mile below any freshwater source (Fig. 4-4). You don't have to be a geologist or an expert in rock physics to appreciate the illogic of nearly imperceptible amounts of a chemical being pumped into a subterranean formation a mile, even two miles, down and somehow quickly percolating back up through thousands of feet of thick granite and other types of rock to taint freshwater supplies closer to the surface.

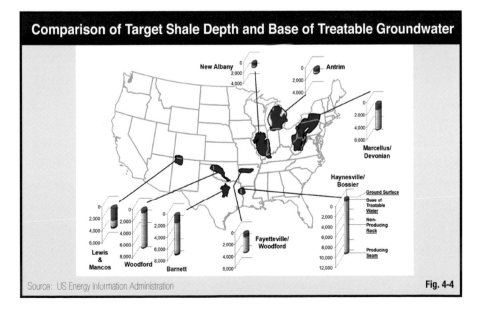

Comparison of Target Shale Depth and Base of Treatable Groundwater

Source: US Energy Information Administration

Fig. 4-4

I previously discussed oil and gas migrating up to the surface and forming seeps, an apt comparison if you overlook the orders-of-magnitude difference in volume of the migrating hydrocarbons. In any event, that process takes millions of years. If the Earth's pressure does cause the chemicals injected deep into the earth today to seep to the surface in millions of years, the same will be true of any number of naturally occurring toxins in those same formations.

The risk isn't really in the deep well injection of these chemicals. The biggest risk of even negligible contamination would come from spilling the container holding one of these additives at the surface. So you might as well shut down all the auto repair shops and swimming pools too, because it's dead certain that commonly used chemicals like antifreeze and pool cleaner get spilled and even dumped down storm drains by the millions of gallons every day, a thousand times more likely to contaminate freshwater supplies.

Trade Secrets Versus Transparency

If you want an idea of the main reason why the oil and gas industry has been fighting an uphill battle against opponents of fracking, I'll remind you about the industry's poor track record of educating the public about pretty much everything it does.

A poll conducted in October 2013 by the University of Texas at Austin found that only 38% of Americans polled support hydraulic fracturing—a drop from 45% six months prior.[62] And yet 57% see domestic natural gas production as beneficial in lowering carbon emissions—an increase from 53%. Then again, that 38% who supported fracking wasn't 38% of all those polled—just that portion of the total 40% of the entire poll population who said they were familiar with the technology. And 48% of those aged fifty-five and older (who are familiar with fracking) support fracking versus 31% of those under fifty-five; however, of the older group, 29% think the US should permit natural gas exports versus 37% of the younger group.

---

62 http://www.utexas.edu/news/2013/10/17/ut-energy-poll-shows-public-disconnect/

Additionally, 58% of those polled think America's largest supplier of oil is Saudi Arabia, while only 13% picked Canada—the correct answer.

The poll results underscore the real disconnect between energy and the American public, according to UT Energy Poll Director Sheril Kirshenbaum.

"In some instances, ideology may influence attitudes, but there's unquestionably a lack of understanding across a broad swath of energy issues that affect each of us," she said.[62]

As an industry, we've done a lousy job explaining what we do and how we do it. Terms such as "trade secrets" and "proprietary process" are used too often, made possible by loopholes in most states' disclosure laws that say if the chemical content of the fluid is unique, you can say it's proprietary and not have to disclose it. In effect, the industry says to John and Jane Doe, who own a farm where there's drilling going on, "There are a whole bunch of chemicals I'm putting down that hole, and they're all top secret."

Instead, the industry should disclose what the chemicals are and explain the safety precautions being taken—the extra layers of pipe, the depth of the pipe, how we know where the water aquifer is, and so on. Then we need to work on removing any scary-sounding chemicals from the mix and making them all naturally occurring, food-grade safe.

There are companies working on that now, including my own company's EnviroFrac™ program. Under the EnviroFrac™ program, any additive not critical to the successful completion of a well is eliminated, and we work to find greener alternatives for all essential additives. To date, we have eliminated 25% of the additives used in our frack fluids. Along with many other companies seeking alternative methods for fracking without using water, our EnviroFrac™ program strives for 100% reuse or recycling of water used in fracking as well.

Despite the so-called loophole provided by the "Halliburton Rule," even Halliburton has devoted great efforts to developing friendlier frack fluids. There's poetic irony in that the most widely publicized acts of demonstrating the irrelevance of the Halliburton loophole involved Halliburton executives drinking their food-grade frack fluid in front of a conference audience.[63]

---

63 http://business.financialpost.com/2013/10/31/haliburton-fracking-fluid/?__lsa=9653-a444

There is currently no federal mandate requiring the disclosure of frack fluids, but many in the industry have recognized the error of trying to keep frack fluid components secret and have begun disclosing their "recipes" anyway.

That's where FracFocus comes in.[64] FracFocus is the national hydraulic fracturing chemical registry. FracFocus is managed by the Groundwater Protection Council and Interstate Oil and Gas Compact Commission, two organizations focused on conservation and environmental protection. The FracFocus website provides public access to the list of chemicals contained in the fluids that are used in fracking operations in the site user's area. It also does a good job of explaining how fracking works, what the chemicals are used for specifically, and what steps the industry takes to protect groundwater. The states are signing on to use FracFocus as a means of official state fracking chemical disclosure, including these ten as of publication: Colorado, Louisiana, Mississippi, Montana, North Dakota, Ohio, Oklahoma, Pennsylvania, Texas, and Utah.

The industry needs to be regulated, but it goes beyond what's in the fracking fluids. We should also be concerned about who is pumping them into the ground. In Texas, for example, anyone can go to the Railroad Commission of Texas, which regulates the oil industry in the state, fill out some forms, put up a $25,000 cash bond, and become a licensed oil and gas operator.

Is that sufficient? Shouldn't operators have to go through some kind of verification or validation process to ascertain whether they know what they're doing and, if they don't, require them to pay another, already qualified and experienced party to operate their wells? It makes basic commonsense to set up some kind of safety training and education process that helps protect the common good while weeding out the potential bad actors.

I'm not a big fan of regulation, but when someone causes environmental damage or is heedless of safety measures, it reflects badly on

---

64 http://fracfocus.org

everyone in the business. There is no specific mechanism for weeding out the careless operators. When the industry is perceived as out of control because of the recklessness of a few bad actors, it puts us all in jeopardy. It makes it harder to do what we do well. Would *Gasland* have had the swift and lasting impact it has had if the BP Gulf of Mexico disaster hadn't happened? I doubt it.

The environmental movement that inspired Josh Fox and has, in turn, been energized by him, has done a lot of good for humanity, but it hasn't always done a good job of acknowledging that every energy choice is a trade-off, or that the world will remain dependent on fossil fuels for many years to come. The cameras and film used to make the movie, the fuel to power Yoko Ono's limousine to deliver her to the New York screening of Fox's sequel, *Gasland 2*, the air conditioning to make the theater comfortable, on and on—all of it depends on afford-able hydrocarbons.

In fact, that's a fraction of the items we rely on daily that are made with petroleum. Did you know that more than half of each barrel of oil consumed in the US is used to manufacture things like heart valves, ink, bicycle tires, clothing, roofing, vitamin capsules, shoes, soap, tape, paint, drinking cups, soft contact lenses, candles, hand lotion, carpet, sporting equipment, fertilizers, lipstick, and about 6,000 more products that fill our homes, get us where we need to go, help us main-tain our health, support our food industry, and provide tools for our work?[65] We are decades away from alternative energy and other sources replacing oil and gas to provide the standard of living to which we've become accustomed.

In my travels I find myself debating ardent environmentalists (I'm an environmentalist, too, but also a realist), and I've learned that you can't get them to acknowledge the environmental and economic trade-offs for all forms of energy. Windmills kill more than a half million birds and bats each year.[66] Solar panels also kill birds and insects, are made with

65 http://www.ranken-energy.com/Products%20from%20Petroleum.htm
66 http://www.smithsonianmag.com/smart-news/how-many-birds-do-wind-turbines-really-kill-180948154/?no-ist

toxic chemicals, and have to be disposed of properly when they wear out.[67] Nobody wants to have those chemicals near them. China has been caught multiple times during the manufacturing process of solar panels dumping chemicals that are so harmful you can't let them touch your skin.[68]

Some environmentalists are against coal because it's dirty, they're against natural gas because of fracking, they're against oil because of spills, and they're against all of them because of greenhouse gases. They will shout all that in my face and then leave the hall where I've spoken to drive home in their hybrid or electric car that's built using many petroleum-based products and powered by toxin-laden batteries that are refueled with electricity generated from burning coal.

Gasland made fracking a dirty word but, done correctly and carefully, it's not the dirty, dangerous process it's been portrayed to be. As with any industrial activity, there are risks in drilling, completing, and producing wells. But the advances in horizontal drilling and hydraulic fracturing are just that—advances in standard techniques that the oil and gas industry has employed safely a million times more than it has not. It's not as if there had been a Manhattan Project that spawned a wholly new source of seemingly limitless energy that also had the potential to be weaponized and threaten all life on Earth, as was the case with nuclear energy.

But because these "tweaks" to largely conventional technologies transformed a massive, long-thought-uneconomic hydrocarbon resource into a viable new source of energy, it truly became a game-changer on the scope of Oppenheimer's "child."

The fracking revolution came on so strong and so fast it caught everyone, including the oil and gas industry, flat-footed. A huge opportunity for collaborating with the environmental movement and the public was squandered. You'd think the big international majors in the energy business would have the budgets and the motivation to do just that, but it turns out their reluctance stems from being easy targets for their critics because of their size and generally aloof nature. ExxonMobil holds its

---

67 http://www.ucsusa.org/clean_energy/our-energy-choices/renewable-energy/
   environmental-impacts-solar-power.html
68 http://www.washingtonpost.com/wp-dyn/content/article/2008/03/08/AR2008030802595.
   html

board and shareholder meetings next to the building where my offices are in Dallas and every meeting brings huge protests condemning the giant company—that they make too much money, they're greedy, they're ruining the world. When I pass those protesters on my way to the office, I wonder, what's your plan?

Gasland and the anti-fracking movement have distracted the public from an inconvenient truth: Fracking has unlocked a future of prosperity and energy security no sane person would have predicted just a few years ago and no reasonable person would reject. After nearly a half-century of our economy being held hostage by foreign oil regimes and volatile world markets, the US is poised to become the world's leading producer of oil and gas and to preside over an incredible renaissance in other industries benefiting from the bounty of low-cost natural gas.

Out of all this cacophony there is one indisputable fact—fracking is here to stay. Attempts to derail the fracking boom will fail, because the train has already left the station. Since fracking started in the 1940s, the process has yielded more than 7 billion barrels of oil and more than 600 trillion cubic feet of natural gas in the US, according to the American Petroleum Institute. API contends that without the advances in fracking and horizontal drilling, America would lose 45% of domestic natural gas production and 17% of her oil production within five years.[69] Five years! How is that not a prescription for economic collapse?

As I've shown in the previous chapter, how it happened is an only-in-America kind of story: a happy confluence of geology, geography, and our own fiercely entrepreneurial culture.

What started here in America is starting to become a global phenomenon. You might not see it, but the energy boom is already reshaping our world.

In the next chapter, we'll delve into the potential energy security and economic benefits that will flow from the fracking revolution to America and assess the debate over its limits and challenges from a resource standpoint.

---

69 http://www.api.org/hydraulicfracturing

Big Caesar #2H was Breitling Energy's first horizontal
Bakken/Three Forks development well.

Chapter 5
Come the Revolution: America

Bonanza or Bust?

The new American energy revolution has arrived. So what's in it for us, as Americans, knowing that we have these huge oil and natural gas resources and that we are capable of economically recovering them in a safe and environmentally friendly manner?

Chapter 2 detailed the scope of what we know about America's overall hydrocarbon resource base. We've known about those kinds of resources for decades, and as our knowledge of them has grown, so too have our estimates of their potential ultimate size.

Identifying a resource and estimating its potential size are interesting—and important exercises. The latter exercise is and always has been a subject for debate and conjecture among energy experts. All too often not enough attention is paid to the parameters that make understanding resource potential useful: the technology and price needed to convert resources to reserves and then to production. In other words:

- What will it take, in terms of technology and price (always intertwined), to monetize this resource at a cost that allows for a decent return on investment while still remaining competitive in a global market?
- What is the likelihood that the economics that underlie this viable monetization will remain attractive in the long

term to the resource developer but also long enough to ensure that the resource fulfills other strategic needs, i.e., becoming a net positive for the US economy, energy security, and environmental goals?

We'll get to those points shortly, but for now let's just take a look at how much of a difference the fracking boom has already made in the US.

Rebuilding Reserves

Purely in terms of US proved oil and natural gas reserves, the impact of the fracking boom has been stunning. For decades, US proved oil and gas reserves have been steadily declining or barely maintaining a static level. Obviously, if you can't find more oil and gas each year than you can produce, your reserves will decline. And we've noted here that the bulk of reserves additions in America in recent years has been reserves growth—extending existing finds—rather than through exploration and discovery.

Advances in fracking and horizontal drilling essentially amount to a "technological discovery" of a resource we've long known was there. And that "discovery" is of a scale we haven't seen in this country since the 1970 discovery and delineation of the Prudhoe Bay oilfield. Figures 5-1a and 5-1b show the big jumps in oil and gas reserves from just the mid-2000s to 2011. Starting in 1979 through 1981—shortly after the addition of Prudhoe Bay and other Alaskan North Slope reserves pushed US total proved oil reserves to 31.3 billion barrels—the US total fell by about one-third, or 10 billion barrels, by 2008. That's when oil production from shale and tight sands really started to ramp up, the decline came to a screeching halt, and the curve began to shoot upward, with the US adding almost 9 billion barrels since 2008.

The US EIA lags somewhat in updating its reserves numbers. But the *Oil & Gas Journal*, which updates its estimates of oil reserves every year, most recently (December 2013) pegged total US oil reserves at 31.8 billion barrels, an increase of almost 3 billion barrels over 2012. In other words, in the space of only five years, largely due to the fracking boom, the US

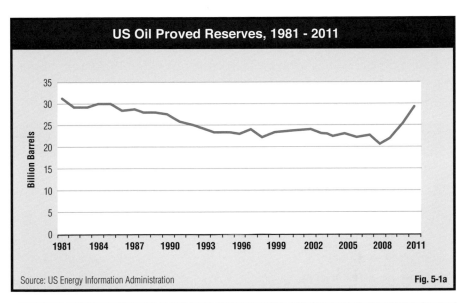

US Oil Proved Reserves, 1981 - 2011

Billion Barrels

Source: US Energy Information Administration

Fig. 5-1a

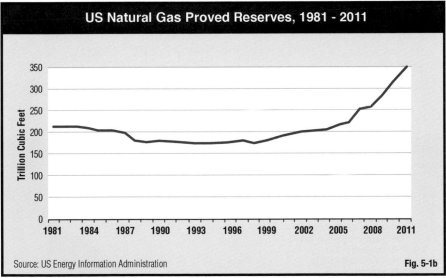

US Natural Gas Proved Reserves, 1981 - 2011

Trillion Cubic Feet

Source: US Energy Information Administration

Fig. 5-1b

has replaced all of the net reserves lost over nearly three decades. That's a stunning achievement by anyone's reckoning.

On the natural gas side, the reversal has been even more profound if not as recently abrupt, as Figure 5-1b also shows. In the early 1980s, US proved natural gas reserves averaged about 208 trillion cubic feet (Tcf) each year. Generally low gas prices—from 1980 to 2000 the average US wellhead gas price was just under $2 per thousand cubic feet (Mcf)—didn't provide enough incentive for operating companies to do much more than running all out just to stay in place. For the rest of the 1980s and through the 1990s, US proved natural gas reserves averaged about 175 Tcf.

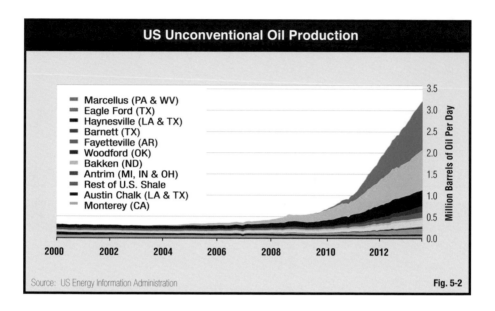

Fig. 5-2

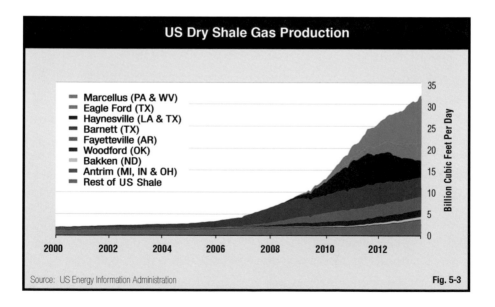

Fig. 5-3

By the early 2000s, the convergence of technology and price, notably in the emerging Barnett Shale, started the upward trend on proved gas reserves growth. Since the early 2000s, US proved natural gas reserves exploded to double the early 1980s level. According to the EIA data, from 2004 to 2011 alone, US proved gas reserves jumped by a staggering 75%, thanks largely to the giant shale plays such as the Barnett, Haynesville, and Marcellus. According to *Oil & Gas Journal*'s year-end 2013 estimates, US proved gas reserves increased by another 30 Tcf, bringing the country to a grand total of 372 Tcf.

It's important to keep in mind the distinction about resources versus reserves that I made earlier. When you hear talk about the shales providing a 100-year supply of gas for the US and then you look at current annual consumption of about 26 Tcf per year versus current proved reserves—well, that's some easy math that works out to only about fourteen years' worth of gas. This is a deceptive exercise that has often been used by those opposed to oil and gas development—and it's based on the false premise that no further additions to reserves occur. Even with the net declines we saw in oil and gas reserves, if we hadn't been adding to reserves while producing from them, we'd have been in a world of hurt a long time ago.

So what about that 100-year claim? Some of the estimates have since been ratcheted down by the US Geological Survey, but the latest estimate for technically recoverable shale gas, tight gas sands, and coalbed methane resources combined is about 1,600 Tcf. So we're taking a reasonable assumption that these recoverable resources will convert into proved reserves in time—and that latter metric can shift up or down according to price—to give us about sixty-two years' worth of gas if consumption were to remain static. Which it won't, of course. But then neither will the resources nor the reserves numbers. We are very early in the unconventional gas production game. According to the EIA, to date about 1–3% of just the technically recoverable shale gas resource has been produced.[70]

---

70 http://www.eia.gov/analysis/studies/usshalegas/pdf/usshaleplays.pdf

Growing Production

As exciting as the growth in US reserves has been, the truly valuable impact of the fracking boom has been the surge in US oil and gas production.

Figures 5-2 and 5-3 show the explosive growth in recent years of US "tight oil" production (from shale, tight sands, and other low-permeability formations) and just the dry shale gas production, both by specific play. The biggest contributors among the plays, are, of course, the ones making the headlines. But it's interesting to note that California's Monterey Shale at the bottom of Figure 5-2 is still so very small—due largely to technical challenges and local opposition to fracking—and yet the Monterey may prove to be the biggest of all shale oil plays at an estimated 15.4 billion barrels of oil in place, according to a study by INTEK for the US Department of Energy.[71] Will California producers overcome the challenges and repeat the mind-boggling success we've seen in North Dakota (Bakken Shale) and Texas (Eagle Ford, Permian Basin)? Time will tell.

The data shown in Figure 5-4 represents the shot heard 'round the world as far as US oil and gas producers are concerned, garnering some of the most rewarding headlines for the industry in many years: In 2013

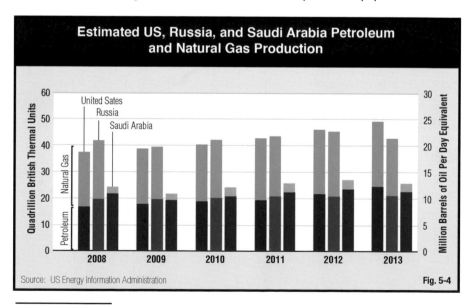

Source: US Energy Information Administration

Fig. 5-4

---

71 http://www.eia.gov/analysis/studies/usshalegas/pdf/usshaleplays.pdf

America surpassed both Russia and Saudi Arabia in terms of combined oil and natural gas production. That pronouncement by the US Department of Energy was greeted as unexpected good news in a year otherwise filled with continuing gloom over a listless economy. And it put Americans on notice that the domestic oil and gas industry was back—in a big way.

Looking at US oil production from a global perspective, Figure 5-5 illustrates how important unconventional oil has become to the US in relation to the rest of the world. In a very short time, the US has grown its market share of world oil production to 10%. Now, 10% may not sound like a commanding market share, especially compared with OPEC's 33% market share, but increasing production of oil from non-OPEC sources always makes it more difficult for the cartel to raise prices, and often the price inflection point is at the margin.

What I mean by that is that OPEC maintains a certain supply cushion—how much it is producing in relation to its total productive capacity—and if that differential shrinks, nervous markets start bidding up crude oil futures prices. If that differential expands because the call on OPEC oil declines as its members maintain or increase their own production, it tends to soften oil prices, or at least diminish the potential for damaging oil price spikes.

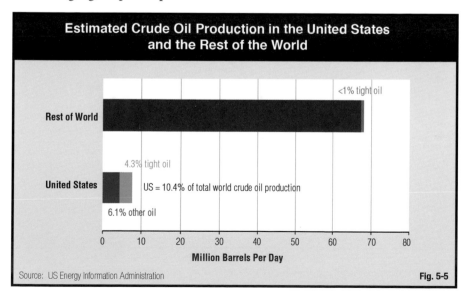

Estimated Crude Oil Production in the United States and the Rest of the World

Source: US Energy Information Administration

Fig. 5-5

OPEC's own Secretariat—which produces a reliable and largely neutral monthly assessment/forecast of oil supply and demand—predicted that the call for OPEC oil would decline in 2014 even as global oil consumption would continue to rebound.

According to the OPEC Secretariat, non-OPEC oil production will expand by 1.1 million barrels per day in 2014, mainly from the US, while the call for OPEC oil will slip by about 300,000 barrels per day, or 2.6%.[72]

How high can America go with oil and gas production? Figure 5-6 offers three scenarios for US crude oil production, with the best case peaking at a record of more than 10 million barrels per day in the next decade—and being sustained there for two more decades. Figure 5-7 breaks out those forecasts by US production area and scenario, with virtually all of the growth coming from Lower 48 tight oil. Obviously, the reference case is the most likely one, taking a conservative view that doesn't rely on innovations in the ultimate recovery of unconventional resources.

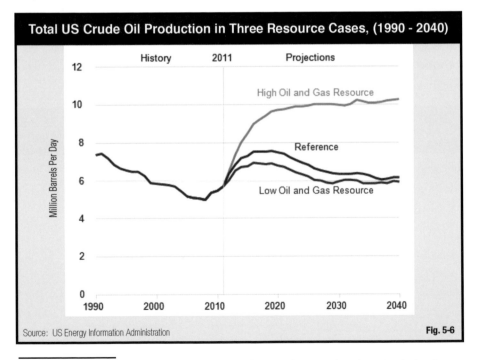

Total US Crude Oil Production in Three Resource Cases, (1990 - 2040)

Source: US Energy Information Administration

Fig. 5-6

---

72 http://business.financialpost.com/2013/07/10/opec-to-lose-market-share-to-shale-oil-in-2014/?__lsa=edce-1b0a

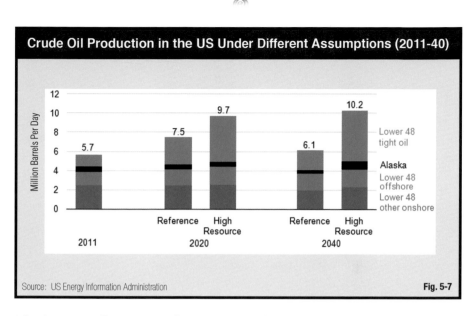

Crude Oil Production in the US Under Different Assumptions (2011-40)

Source: US Energy Information Administration

Fig. 5-7

That's especially true at the juncture of unconventional resources with enhanced oil recovery processes that have been successfully used for decades (see Chapter 2) but never before applied to these newly fracked oil and gas shales and tight sands plays.

In any event, EIA acknowledges these forecasts are always moving targets. Just take a look at the agency's grim outlook for US oil production way back in 2000 (Fig. 5-8).

And yet, some of my colleagues have gotten a bit overly enthusiastic with some of this "Saudi America" talk. As Figure 5-9 shows, the US will almost certainly still be importing a net one-third of its supply of oil and other liquid fuels twenty-five years from now. But that's about half of US oil import dependency as recently as 2005.

As great as the fracking boom has been for boosting US oil reserves and production, it's important to keep things in perspective. Even tossing around numbers such as 400 billion barrels of discovered, technically recoverable conventional oil and 30–50 billion barrels of identified unconventional oil resources for the US, it's apples-to-oranges when you compare US oil potential with Saudi Arabia's. Remember, converting those resources into proved reserves will take a nearly unthinkable sum of capital, an oil price floor of at least $80, and further technology advances.

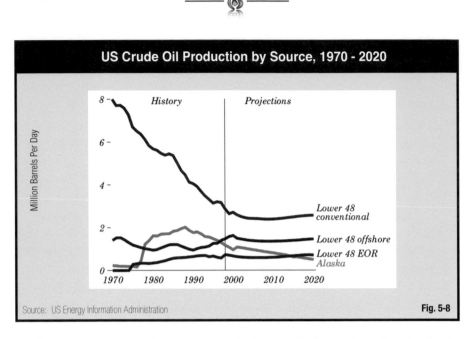

US Crude Oil Production by Source, 1970 - 2020

Million Barrels Per Day

History — Projections

Lower 48 conventional
Lower 48 offshore
Lower 48 EOR
Alaska

1970 1980 1990 2000 2010 2020

Source: US Energy Information Administration

Fig. 5-8

Our current proved reserves total about 32 billion barrels. No doubt that number will rise, but even at the 2012–2013 rate of growth of 3 billion barrels, it would take decades for America to reach a proved reserves level of 266 billion barrels—the estimated total proved conventional reserves in Saudi Arabia. There's some jockeying among Saudi Arabia and Venezuela (298 billion barrels, according to *Oil & Gas Journal*) for the title of number one in terms of proved oil reserves. However, depending on which source you use, as much as one-third of Venezuela's touted 298 billion barrels is ultra-heavy (very viscous) crude that is technically recoverable but requires a commitment of new technology, high oil prices, and gigantic capital sums to be economic. This stuff would probably cost more to develop than the Canadian oil sands or US shales.

Now consider Saudi Arabia's situation: Essentially all of its proved oil reserves are conventional in nature. The lion's share of its oil reserves and production resides in a mere eight oil fields. One such field, Ghawar, alone produced 5 million barrels of light, sweet crude oil per day, and it still has at least 70 billion barrels of oil remaining.

So, yes, the US produces more "oil" than Saudi Arabia at the moment, including products such as natural gas liquids and other non-crude

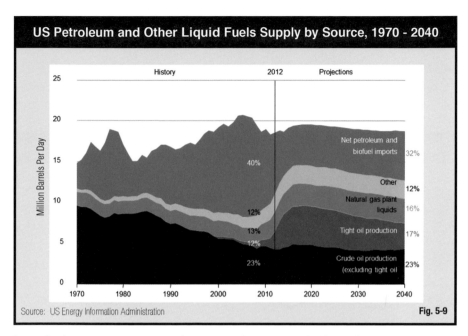

US Petroleum and Other Liquid Fuels Supply by Source, 1970 - 2040

History   2012   Projections

Million Barrels Per Day

25

20

15

10

5

0

1970   1980   1990   2000   2010   2020   2030   2040

Net petroleum and biofuel imports 32%

40%

Other 12%

Natural gas plant liquids 16%

12%

Tight oil production 17%

13%

12%

Crude oil production (excluding tight oil 23%

23%

Source: US Energy Information Administration   **Fig. 5-9**

liquids, but comparing crude-to-crude output, the kingdom still outpro-duces the US at 10 million barrels per day to 8.5 million. Saudi Arabia's productive capacity is another matter altogether. It has the capability to ramp up crude production from about 10 million to 12 million barrels per day within ninety days. The kingdom also has toyed with the idea of increasing that capacity to 15 million barrels per day, but the US fracking boom, together with rising deepwater production from other non-OPEC nations, seems to have quashed that talk for now.

More importantly, Saudi Arabia enjoys a lifting cost of only about $5 per barrel. Compare that with lifting costs of about $20–$30 per barrel for conventional oil in the US in recent years. However, it's been speculated that the breakeven point for hugely expensive oil shale plays is closer to $60 per barrel. If oil prices collapsed to, say, $50 per barrel, much of US and Canadian oil production would halt and be shut in. When my colleagues in the industry say, "We're not running out of oil; we've run out of cheap oil," they are thinking about North America, not Saudi Arabia.

No doubt, America's energy renaissance is a remarkable turnaround in what until recently was deemed a sunset industry. While the fracking

boom holds and oil prices remain buoyant, the US can—and should—continue to enhance its energy security and bolster its economy with surging oil production.

But it is not Saudi Arabia and never will be.

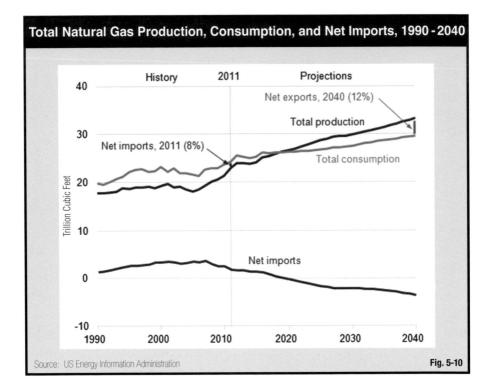

Total Natural Gas Production, Consumption, and Net Imports, 1990-2040

Source: US Energy Information Administration

Fig. 5-10

Natural Gas: Shale Bonanza

If not "Saudi America," could the US instead become "Russia America," at least in terms of natural gas? Certainly America's dry gas production has topped Russia's for more than two decades now. When it was still leading the USSR, Russia and the other Soviet republics combined comprised the number one natural gas producer, a title relinquished to the US with the collapse of the Soviet Union. Even at that, Russia still holds by far the world's largest proved conventional dry gas reserves at about 1,600 Tcf,

more than quadruple the same metric for the US and roughly equal to America's total estimated *recoverable resources* from unconventional plays. So, as with Saudi Arabia, the comparison isn't apt.

That does nothing to diminish the astounding turnaround that has happened in the US natural gas industry. Figure 5-10 shows the US EIA's most recent forecast that America will become a net exporter of natural gas by 2020, and Fig 5-11 breaks that forecast out by export and import source. Again, that's net exporter. We've been exporting gas to Canada and to Mexico via pipeline for decades now, while importing some from both neighbors at varying levels, along with imports of liquefied natural gas (LNG) from a wide range of countries.

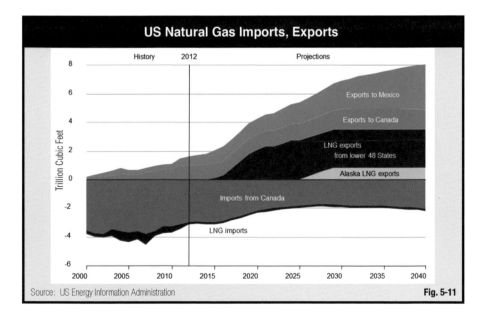

US Natural Gas Imports, Exports

History   2012   Projections

Exports to Mexico
Exports to Canada
LNG exports from lower 48 States
Alaska LNG exports
Imports from Canada
LNG imports

Trillion Cubic Feet

Source: US Energy Information Administration   Fig. 5-11

The LNG 180°

Not many folks realize that until recently we also exported LNG from Alaska to Japan, a trade that dates back to the 1960s. Back then we were the sole supplier of LNG to Japan. When the Alaska LNG project owners opted to not renew their export license in 2013, it was largely because they didn't see local production (Alaska's Cook Inlet oil and gas fields)

Proposed North American LNG Export Facilities, 2014

Proposed US LNG Export Sites

- Freeport, TX: Freeport LNG Development/Freeport LNG expansion/ FLNG Liquefaction, 1.8 Bcfd
- Corpus Christi, TX: Cheniere Energy (Corpus Christi LNG), 2.1 Bcfd
- Coos Bay, OR: Jordan Cove Energy Project, 0.9 Bcfd
- Lake Charles, LA: Southern Union (Trunkline LNG), 2.2 Bcfd
- Hackberry, LA: Sempra (Cameron LNG), 1.7 Bcfd
- Cove Point, MD: Dominion (Cove Point LNG), 0.82 Bcfd
- Astoria, OR: Oregon LNG, 1.25 Bcfd
- Lavaca Bay, TX: Excelerate Liquefaction, 1.38 Bcfd
- Elba Island, GA: Southern LNG Company, 0.35 Bcfd
- Sabine Pass, LA: Sabine Pass Liquefaction, 1.4 Bcfd
- Lake Charles, LA: Magnolia LNG, 1.07 Bcfd
- Plaquemines Parish, LA: CE FLNG, 1.07 Bcfd
- Sabine Pass, TX: ExxonMobil (Golden Pass), 2.1 Bcfd

Proposed Canadian LNG Export Sites

- Kitimat, BC: Apache Canada Ltd., 1.28 Bcfd
- Douglas Island, BC: BC LNG Export Cooperative, 0.23 Bcfd
- Kitimat, BC: LNG Canada, 3.23 Bcfd

Source: US Energy Information Administration   Table 5-1

Proposed, Existing North American LNG Import Facilities, 2006

US Import Facilities

- Existing
  - **Everett, MA**: Tractebel (DOMAC), 1.035 Bcfd
  - **Cove Point, MD**: Dominion (Cove Point LNG), 1.0 Bcfd
  - **Elba Island, GA**: El Paso Energy (Southern LNG), 0.68 Bcfd
  - **Lake Charles, LA**: Southern Union (Trunkline LNG), 1.2 Bcfd
  - **Gulf of Mexico**: Excelerate Energy (Gulf Gateway Energy Bridge), 0.5 Bcfd

- Approved
  - **Lake Charles, LA**: Southern Union (Trunkline LNG), 0.6 Bcfd
  - **Hackberry, LA**: Sempra Energy, 1.5 Bcfd
  - **Bahamas**: AES Ocean Express, 0.84 Bcfd*
  - **Bahamas**: Tractebel (Calypso), 0.83 Bcfd*
  - **Freeport, TX**: Cheniere (Freeport LNG Dev.), 1.5 Bcfd
  - **Sabine, LA**: Cheniere LNG, 2.6 Bcfd
  - **Elba Island, GA**: El Paso Energy (Southern LNG), 0.54 Bcfd
  - **Corpus Christi, TX**: Cheniere LNG, 2.6 Bcfd
  - **Corpus Christi, TX**: ExxonMobil (Vista del Sol), 1.0 Bcfd
  - **Fall River, MA**: Hess LNG (Weaver's Cove Energy), 0.8 Bcfd
  - **Sabine, TX**: ExxonMobil (Golden Pass), 1.0 Bcfd
  - **Corpus Christi, TX**: Occidental (Ingleside Energy), 1.0 Bcfd
  - **Port Pelican (deepwater Gulf of Mexico)**: ChevronTexaco, 1.6 Bcfd
  - **Louisiana offshore**: Shell (Gulf Landing), 1.0 Bcfd

- Proposed
  - **Long Beach, CA**: Mitsubishi/ConocoPhillips (Sound Energy Solutions, 0.7 Bcfd
  - **Logan Township, NJ**: BP (Crown Landing LNG), 1.2 Bcfd
  - **Bahamas**: El Paso/FPL (Seafarer), 0.5 Bcfd
  - **Port Arthur, TX**: Sempra, 1.5 Bcfd
  - **Cove Point, MD**: Dominion, 0.8 Bcfd
  - **Long Island Sound, NY**: TransCanada/Shell (Broadwater Energy, 1.0 Bcfd
  - **Pascagoula , MS**: ChevronTexaco (Casotte Landing), 1.3 Bcfd
  - **Bradwood, OR**: Northern Star Natural Gas, 1.0 Bcfd
  - **Pascagoula, MS**: Gulf Energy LNG, 1.0 Bcfd
  - **Cameron, LA**: Cheniere LNG (Creole Trail LNG), 3.3 Bcfd
  - **Port Lavaca, TX**: Calhoun LNG (Gulf Coast LNG), 1.0 Bcfd
  - **Freeport, TX**: Cheniere/Freeport LNG (expansion), 2.5 Bcfd
  - **Sabine, LA**: Cheniere LNG (expansion), 1.4 Bcfd
  - **California Offshore**: BHP Billiton (Cabrillo Port), 1.5 Bcfd
  - **S. California Offshore**: Crystal Energy, 0.5 Bcfd
  - **Louisiana Offshore**: McMoRan (MainPass), 1.0 Bcfd
  - **Gulf of Mexico**: ConocoPhillips (Compass Port), 1.0 Bcfd
  - **Gulf of Mexico**: ConocoPhillips (Beacon Port), 1.5 Bcfd
  - **Offshore Boston, MA**: Tractebel (Neptune LNG), 0.4 Bcfd
  - **Offshore Boston, MA**: Excelerate Energy (Northeast Gateway), 0.8 Bcfd

Canadian Approved Terminals

- **St. John, NB**: Irving Oil (Canaport), 1.0 Bcfd
- **Point Tupper, NS**: Anadarko (Bear Head LNG), 1.0 Bcfd

Mexican Approved Terminals

- **Altamira, Tamaulipas**: Shell/Total/Mitsui, 0.7 Bcfd
- **Baja California**: Sempra, 1.0 Bcfd
- **Baja California offshore**: ChevronTexaco, 1.4 Bcfd

*Related US pipeline approved; LNG import terminal pending in Bahamas.
Source: US Energy Information Administration   Table 5-2

able to sustain future LNG supply commitments to Japan, where their market share had dropped to 0.5%. Japan's subsequent detour from an aggressive nuclear power build-out in the wake of the Fukushima nuclear reactor disaster has revived talk of a new Alaskan LNG export facility using gas from the state's North Slope, but the hurdles are so daunting for such a megaproject costing perhaps north of $100 billion that it seems out of the realm of possibility.

What is a very real possibility for a US LNG export facility is one based on America's burgeoning new supplies of shale gas. Not only a real possibility, but already happening. Cheniere Energy is building an LNG export terminal at Sabine Pass, Louisiana, that is expected to start operation in 2015. Recently, the Department of Energy gave approval to the seventh such export facility, while many more such projects are planned in both the US and Canada. (Table 5-1). While it's remarkable that we are building LNG export facilities, it's even more stunning in light of what the prospects were for US LNG just a few years ago. That's before the shale gas boom really took off. At the time, anxieties over being able to source enough gas overseas to feed America's rapidly growing consumption of natural gas had spawned more than sixty LNG import facilities that later shook out to thirty-nine, in addition to the five active import terminals we already had in 2006 (Table 5-2).

Flash forward from 2006, and it's apparent that the US LNG business has done a complete 180° and will again be in the LNG export business by 2015. Table 5-1 shows only the sixteen formally proposed LNG export facilities in North America; there are additional potential LNG export terminal sites that have been identified by project sponsors.

Even before a single new LNG export facility could come online, events involving Russia and Ukraine in early 2014 suddenly thrust the US energy boom onto the global geopolitical stage and our new natural gas bonanza has become a strategic asset for the US and its allies. More about that in the next chapter.

The Boom's Economic Benefits

OK, so the fracking boom is great for US energy security and trade, but how significant is it in terms of benefits to the American economy?

It's difficult to overstate the benefits that are flowing from our unconventional oil and gas bonanza already and in the years to come. By 2020, says the *Economist*, the fracking revolution will have added 2–4% ($380–$690 billion) to American GDP and create more than twice as many jobs as automaking provides today.[73] That's truly mind-boggling. US GDP today is about $16 trillion, and US automakers employ about 800,000 people.

The numbers keep getting better. IHS, a Colorado-based information and data services firm, conducted studies in 2012 and 2013 that found that the economic benefits from the fracking boom include the following:[74]

- Growing US oil and gas industry jobs by 40% from 2007–2012, compared with a 1% increase in the rest of the private sector during the same period.
- Supporting 2.1 million jobs in 2012, growing to 2.5 million in 2015, 3 million in 2020, and 3.9 million in 2035.
- Of the jobs indirectly related to fracking, adding 515,000 manufacturing jobs by 2025. Petrochemicals employment alone will increase from 53,000 in 2013 to 149,000 in 2015 and almost 319,000 in 2025.
- Increasing government revenues (federal, state, local) by $1.6 trillion from 2012 to 2025.
- Lowering the US trade deficit by $180 billion by 2022.
- Spending $216 billion on infrastructure from 2012 to 2025.
- Adding to personal per capita wealth by $1,200 in 2012 and $3,500 in 2025, mainly through energy savings and income increases.

---

73 http://www.economist.com/news/leaders/21596521-energy-boom-good-america-and-world-it-would-be-nice-if-barack-obama-helped
74 http://www.energyxxi.org/sites/default/files/pdf/americas_new_energy_future-unconventional_oil_and_gas.pdf

Turns out the energy renaissance is also spurring a manufacturing renaissance as well, thanks to a cost advantage over foreign competitors of 5–25% that American plants and factories will be enjoying by 2015 (Fig. 5-12).

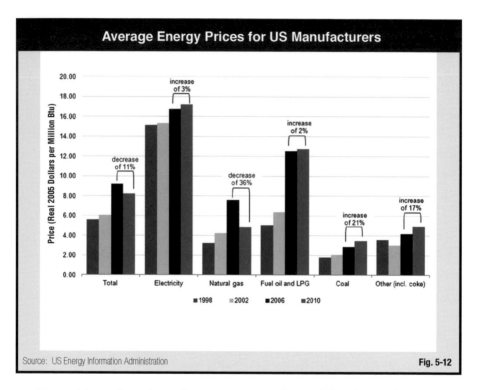

Since 2011, plans have been announced to add industrial manufacturing facilities totaling 128 new plants, mostly in petrochemicals, along the Gulf Coast alone, where plastics exports are spiking even as plastic resin producers are enjoying significant new energy savings (Fig. 5-13).[75] Fertilizer, cement, tire, aluminum, iron, and steel companies nationwide are also adding capacity, thanks to the lower gas prices the fracking boom has unleashed.

Are we seeing a New American Industrial Revolution?

Of course, while the nation benefits from the fracking boom in many

---

75 http://www.economist.com/news/business/21589870-capitalists-not-just-greens-are-now-questioning-how-significant-benefits-shale-gas-and

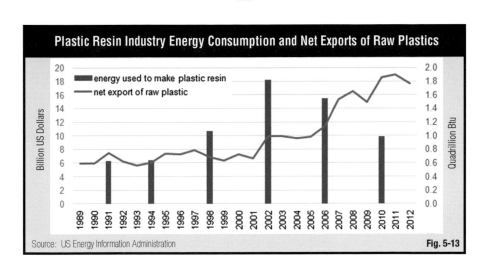

Plastic Resin Industry Energy Consumption and Net Exports of Raw Plastics

- energy used to make plastic resin
- net export of raw plastic

Source: US Energy Information Administration

Fig. 5-13

ways, the economic impacts at the local and regional level, especially in the boom states, have been truly transformative.

Much has been said about how the shale boom has turned North Dakota—a minor oil producer—into the equivalent of an OPEC producer; its oil production has rocketed from less than 100,000 barrels per day ten years ago to closing in on 1 million barrels per day. The result: The state has the nation's fastest-growing economy and lowest unemployment rate at 3%.

Until the explosive growth of the Eagle Ford play a few years ago, South Texas was a long-depressed region of about 1.1 million people. The University of Texas at San Antonio recently conducted a study that found that the Eagle Ford boom:

- Generated $61 billion and 116,000 jobs for the twenty-county region in 2012
- Will generate $89 billion and 127,000 jobs for the twenty-county region in 2022
- Added more than $1 billion in total local government revenue in 2012
- Provided $1.2 billion in estimated state revenue in 2012

Even states long dependent on the coal industry are gaining new opportunities from the very resource that is displacing coal: shale gas. The future

hasn't been looking very bright for the US coal industry lately, given the acceleration of coal-fired power plant retirements looming on the horizon (Chapter 1, Fig. 1-15). After peaking in 2008, coal production by 2013 had fallen 80% in Pennsylvania, 50% in Ohio, and 32% in West Virginia. Even with the economic woes emanating from coal's struggles, officials in those states look to the booming Marcellus and Utica shale plays as the best hope of replacing lost coal jobs—through shale gas industry gains.[76]

The shales taketh away and the shales giveth? More like the horse-drawn buggy makers are going to work for Henry Ford.

The Shale Bubble?

Of course, whenever there's a compelling reason to develop and produce more oil and gas, there are skeptics about the long-term viability of the resource, as we saw with the peak oil contingent in Chapter 1.

That's another debate that has raged around fracking and unconventional resources. Beyond the environmental concerns they raised, opponents of fracking have also made claims about the long-term viability of shale oil and gas and other unconventional hydrocarbons. Essentially, they contend that the fracking frenzy and the shale boom essentially constitutes a "bubble" that's about to burst, as we saw with the dot-com and real estate cycles. The first sign of such a bust is that investors are pumping more money into an asset than it's really worth because of their overinflated expectations of ever-growing value.

An analysis by the environmentalist think tank the Post Carbon Institute concluded that shale gas and tight oil (including shale oil) can't fulfill our "exuberant" expectations and the current "bubble" will last only about ten years.[77]

The February 2013 study noted that 80% of shale gas production comes from five plays, several of which are in decline. Additionally, the very high decline rates of shale gas wells require something on the order of $42

76 http://www.forbes.com/sites/kensilverstein/2013/11/14/coal-dependent-states-get-
   second-chance-with-shale-gas-boom/
77 http://www.postcarbon.org/drill-baby-drill/

billion spent each year to drill more than 7,000 wells in order to maintain production. Then the study author pointed out that the value of shale gas produced is just $32.5 billion. He also claimed that the best shale gas plays, such as the Haynesville Shale, are already in decline. He did the same exercise for tight oil/shale oil: $35 billion per year to drill 6,000 wells just to maintain production. He contends that tight oil production will peak at 2.3 million barrels per day in 2017 and that all drilling locations in the Bakken and Eagle Ford will have been used up.

So does that mean that all of this drilling activity is a money-losing proposition and won't meet production forecasts? Well, first, let me point out that the study author's executive summary touting these values conveniently omitted the fact that maintaining roughly 3 million barrels per day of current tight oil production (Fig. 5-2) at 2014 oil prices represents a value of about $110 billion per year. Not too shabby a return, and pretty good refutation of the author's prediction of a 2.3 million barrel per day peak in three years, since profits drive further production.

And while Haynesville gas production has declined sharply over the past couple of years, there's a reason for that that has nothing to do with its ultimate potential: Low gas prices crimped drilling. From a peak of about 100 wells per month in March 2010, operators were drilling about 10 wells per month by March 2012.[78] The play once hosted more than 200 rigs; now it's about 50. Fewer wells means less production. Imagine that.

But even with the overall decline in Haynesville gas production, operators are improving recovery rates. Figure 5-14 shows that new-well gas production per rig in the Haynesville, having quintupled since 2007, is still on the upswing.

It's also important to note that many, if not most, of the shale gas wells being drilled today in plays such as the Eagle Ford, Marcellus, and Utica are in fact not being drilled for their natural gas but for their associated condensate and natural gas liquids. They may still be classified as shale gas wells because natural gas (methane) comprises a preponderance of

---

78 http://www.ogj.com/articles/uogr/print/volume-1/issue-4/eagle-ford/haynesville-continues-decline-as-operators-seek-out-wet-gas-plays.html

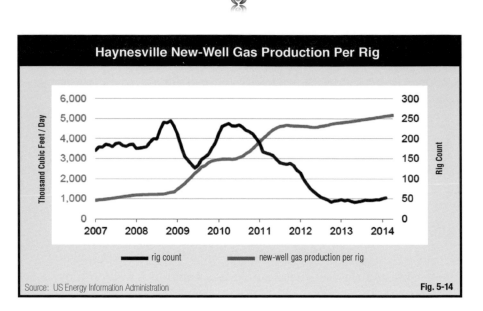

Haynesville New-Well Gas Production Per Rig

Legend: rig count — new-well gas production per rig

Source: US Energy Information Administration   Fig. 5-14

the total hydrocarbon well stream, but the far greater value being realized from that well stream is from its constituent liquids. So to correlate the value of dry shale gas produced with the capital expended on all shale gas wells is misleading.

But there is no disputing the high decline rate of shale wells. Figure 5-15 depicts the decline curve of a typical well in the Barnett Shale: Initial production (IP) is about 2.5 million cubic feet per day, and the first-year decline is a whopping 70% before leveling off to a rate less than one-fifth of the IP rate. But then it continues to produce at that reduced rate for many years, at little incremental operating cost. This is actually one of the major appeals of investing in the fracking boom: Early payouts due to the huge IP rates, and even the reduced production, are essentially all gravy for a long time.

Take for example the Bakken Shale. A typical Bakken well costs about $9–10 million to drill and complete (frack), according to Motley Fool contributor Matt DiLallo.[79] And the typical Bakken well will produce for forty-five years. He adds: "Over those forty-five years the average Bakken well will produce around 665,000 barrels of oil. The economics are excep-

79 http://www.fool.com/investing/general/2013/05/15/10-incredible-numbers-from-the-bakken.aspx

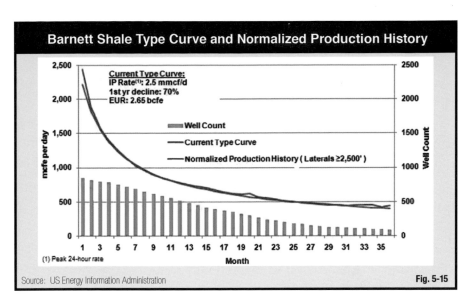

Barnett Shale Type Curve and Normalized Production History

Current Type Curve:
IP Rate[(1)]: 2.5 mmcf/d
1st yr decline: 70%
EUR: 2.65 bcfe

Well Count
Current Type Curve
Normalized Production History (Laterals ≥2,500')

(1) Peak 24-hour rate

Month

Source: US Energy Information Administration

Fig. 5-15

tional even [at] $10 million well costs. For example, if oil drops to $75 a barrel, then the company would still enjoy an internal rate of return of 35% and a payout of just thirty months."

As Continental Resources President Rick Bott said of the Bakken Shale recently, "There is a steep decline, but then there is a very, very long tail. You get flush production, but it's the tail you really count on."[80]

That's a key point to consider. If you drill 5,000 wells in a field, each one may have a high IP rate and then quickly decline into that "very, very long tail" of production until it's no longer economic even at a minimal investment effort. But by then you have 5,000 "tails" stacked atop each other to sustain production at a pretty high plateau for many years.

So what about that claim that all the Bakken locations will be drilled up by 2017? At the time the Post Carbon Institute conducted its study in 2012, Continental Resources estimated the Bakken would be fully developed with 50,000 wells. Continental CEO Harold Hamm recently doubled that estimate.[81]

Michael Lynch, the most prominent opponent of the peak oil theory advocates, exposes the red herring in the decline curve question:

---

80 http://fuelfix.com/blog/2014/03/04/continental-exec-decades-of-life-left-in-bakken/
81 http://www.cnbc.com/id/101466017

But decline rates are only one variable determining production levels, the other being additional drilling. Production is the result of the natural decline rate and offsetting investment, such as infill drilling, water or gas injection, and many other techniques.

There's that pesky reserves growth again. Primary recovery rates in the unconventional plays are pretty low compared with conventional oil and gas plays. The typical recovery factor for a mature conventional oil reservoir is about 20–40%, and for a conventional gas reservoir as much as 80–90%. Compare that with an average recovery factor for shale oil at 3–7% of the original oil in place or 20–40% for shale gas.[82]

But move beyond primary recovery in a conventional oil reservoir to incorporate enhanced oil recovery (EOR) techniques, as discussed in Chapter 2, and it's a new game. Applying EOR processes in conventional oil reservoirs can boost the ultimate recovery factor to as much as 70%. Industry is only just now beginning to look at how EOR techniques might be applied to shale oil reservoirs. What if a technique such as carbon dioxide flooding—so successful in the Permian Basin—could merely double the oil recovery factor in the Bakken or Eagle Ford?

Speaking of curves, experienced operators in the unconventional oil and gas plays have come up substantially on the fracking learning curve, too. Fracking is not typically "one and done." Even before the unconventional fracking boom, it was commonplace to re-enter a well to re-frack it, stimulating production anew.

Re-fracks. EOR. Stacking tails. Maybe when it's all said and done, that notorious downward curve for shale wells will actually amount to a long-lived, healthy production plateau.

And America will enjoy the benefits that flow from the fracking bonanza, not endure a bust.

---

82 http://www.eia.gov/analysis/studies/worldshalegas/

Rig drilling in the Bakken Shale for Breitling Energy.

Chapter 6
Come the Revolution:
The Rest of the World

A World Hungry for Oil and Gas

The fracking revolution may have been born, nurtured, and grown in the US, but it has the potential to become even bigger as a global phenomenon.

Unfortunately, potential does not equate to probability. How rapidly this revolution spreads to the rest of the world could change global oil and gas markets even more profoundly than what has already occurred in the US. There is little doubt that the future need for ample supplies of oil and gas will continue for the foreseeable future.

Despite climate change-related concerns over curbing fossil fuel consumption, expectations are for global oil and gas consumption to continue to grow and together dominate the energy mix. As Figure 6-1 shows, petroleum liquids will still be the largest component of world-wide energy consumption even twenty-five years from now, according to the US EIA. Oil and natural gas taken together will still account for more than half (52%) of global energy consumption by 2040. While renewable energy will be the fastest-growing source of energy, it is not expected to exceed 15% of total energy consumption even twenty-five years from now. That figure includes hydro, with its own environmental baggage (as noted in Chapter 1), which accounts for about half of the electricity

produced by renewable sources. So the idealized notion that solar, wind, and other non-hydro energy sources will carry the load needed to keep the lights on for the planet's 9 billion people in 2040 is pure wishful thinking that runs afoul of all the expert forecasts. I can use other annually updated forecasts of energy consumption (International Energy Agency, OPEC, BP, ExxonMobil, etc.), but they all suggest the same thing: Fossil energy is not being phased out any time soon.

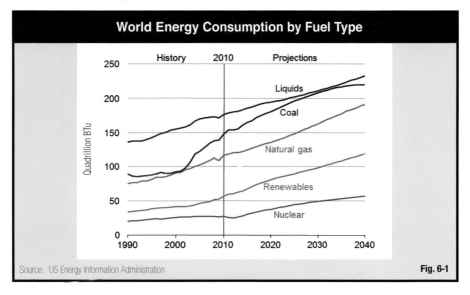

World Energy Consumption by Fuel Type

Source: US Energy Information Administration

Fig. 6-1

There is one caveat about the forecast shown in Figure 6-1, however. It shows coal's market share declining from about 29% today to 27% in 2040, under a scenario that omits curbs on coal consumption to reduce greenhouse gas emissions. Even with its expected rapid growth, natural gas will see its market share grow from 22% to 23% of global energy consumption in the next twenty-five years.

Those shifts will occur almost entirely in the electricity sector. But tough policies to reduce greenhouse gas emissions could change the outlook for both dramatically.

As expected, developing nations will account for the biggest part of future worldwide natural gas consumption (Fig. 6-2). Those nations will almost double their consumption of natural gas by 2040, while developed nations' consumption will rise by less than half.

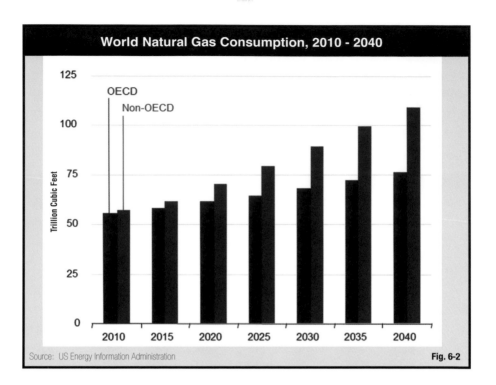

World Natural Gas Consumption, 2010 - 2040

Source: US Energy Information Administration

Fig. 6-2

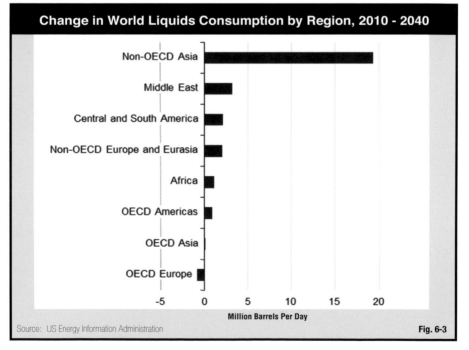

Change in World Liquids Consumption by Region, 2010 - 2040

Source: US Energy Information Administration

Fig. 6-3

But the most dramatic change will come in the form of consumption of liquid petroleum fuels, as developing nations—mainly in Asia—expand the oil market by 28 million barrels per day over the next twenty-five years, while developed nations' oil demand flatlines (Fig. 6-3).

Global Shale Resources

We've already seen how unconventional resources made economic by fracking and horizontal drilling are benefiting the US. With all the coming growth in oil and consumption as developing nations transition from poverty to an emergent middle class—especially given their huge populations—will the global unconventional resources provide a similar benefit?

There is no doubt that global shale oil and gas resources are immense, and in some countries larger than what we find in the US.

Virginia-based consultancy Advanced Resources International (ARI) conducted an assessment of global shale oil and gas resources for countries other than the US for the US EIA in 2011 (updated in 2013). Figure 6-4 shows the distribution of those resources by country, with ARI providing resource estimates for some countries (Tables 6-1a and 6-1b).

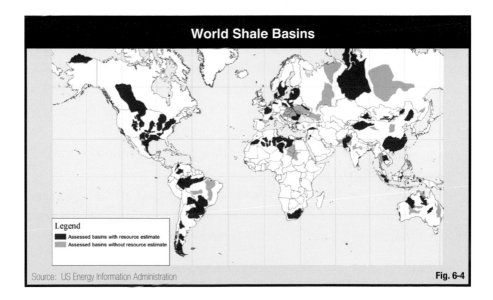

World Shale Basins

Legend
■ Assessed basins with resource estimate
■ Assessed basins without resource estimate

Source: US Energy Information Administration

Fig. 6-4

Table 6-1a: Top 10 Countries Shale OIL Resources			
Rank	Country	Shale oil (billion of barrels)	
1	Russia	75	
2	US	58	48 *
3	China	32	
4	Argentina	27	
5	Libya	26	
6	Australia	18	
7	Venezuela	13	
8	Mexico	13	
9	Pakistan	9	
10	Canada	9	
World Total		345	335 *

Table 6-1b: Top 10 Countries Shale GAS Resources			
Rank	Country	Shale gas (trillion cub. ft.)	
1	China	1,115	
2	Argentina	802	
3	Algeria	707	
4	US	665	1,161 *
5	Canada	573	
6	Mexico	545	
7	Australia	437	
8	South Africa	390	
9	Russia	285	
10	Brazil	245	
World Total		7,299	7,795 *

* EIA estimates used for ranking order. ARI estimates in italics

In the aggregate, 345 billion barrels of oil from shales and even 7,300 Tcf of natural gas from shale are not going to have the same disproportionate, revolutionary impact on the total world oil and gas reserves picture that was the case for the US and its shale oil and gas. In the US, the impact was several orders of magnitude in both cases in terms of recoverable resources. Worldwide, the impact for oil would be akin to finding another Venezuela or Saudi Arabia; for gas, it would be roughly double current proved reserves.[83]

But, as with the US, the impact could be radical at the individual country level. Nowhere is this more the case than in China, which since 2010 has topped the US as the world's largest consumer of energy.

---

83 http://www.ogj.com/articles/print/volume-111/issue-12/special-report-worldwide-report/worldwide-reserves-oil-production-post-modest-rise.html

China's Energy Dilemma

China has been scrambling to nail down oil and natural gas resources around the world for some time, but the urgency now is keen.

The world's most populous nation became the world's biggest producer of energy in 2007 largely on the explosive growth of its coal production to feed the world's fastest-growing major economy. Although it's gradually shifting from coal to natural gas to reduce pollution from burning coal (notably to fuel power plants) China remains wedded to coal, which produces pollution so severe that half the times I fly to Beijing the planes have to be diverted because the smog is too dense to land safely. And the country remains the world's largest emitter of the principal greenhouse gas, carbon dioxide.

Even with its own fairly significant oil and natural gas production, China imports about 60% of its oil and became a net importer of natural gas in 2007.[84]

In my travels there, speaking to audiences of government officials, experts, and oil and gas people, it's clear the Chinese are trying just about everything to secure future energy supplies. They want to know about fracking because China sits on the world's largest technically recoverable reserves of shale gas and the third-largest amount of shale oil.

Over a three-year span I've made a number of speeches in China talking about the technical and practical aspects of fracturing shale. My message has been that fracking for shale gas and oil is not going to help China anytime soon. Topography is a huge challenge. Unlike the US, which has broad stretches of relatively flat land where much of its shale wealth resides, China's shale bounty is either in mountainous areas, densely populated zones, or remote deserts. Its shale deposits tend to be more highly faulted (basically scrambled subsurface geology in which the deposits are less homogenous and broken up into smaller "pockets") than ours.

China also lacks the know-how and equipment to undertake the kind of advanced horizontal drilling and multistage fracturing that we've

---

84 http://www.eia.gov/countries/cab.cfm?fips=ch

perfected in the US and that is critical for transitioning its massive shale bounty from potential resources to production.

China additionally faces water availability challenges. Its growing water shortage is already a huge problem, and fracking requires a tremendous amount of it. China also lacks much of the infrastructure—the roads, the pipelines, the rail lines, the storage facilities—in these shale areas that we take for granted.

The Chinese are proceeding cautiously, so even if the government were to make fracking a priority, it'll be a long time before production makes a dent in the country's energy appetite.

What I've found is that the Chinese don't want to hear about the challenges. A lot of the cautionary comments I have made about its near-term potential never make it out of the room where I'm speaking and into the press. When you read about fracking in China, it's all roses and puppy dogs. They claim they're going to have 60 to 100 billion cubic meters of production coming out of the ground by 2020, which my experience teaches me is impossible and which the EIA has said is "probably unachievable."[85]

As for China's investment in fracking, in 2013 the Chinese had just drilled their sixty-fourth well with no commercial production. In the US, we've drilled more than 50,000, and we have the largest network of oil and natural gas pipelines of any country in the world—spanning nearly 2.6 million miles. China's pipeline infrastructure is woefully inadequate at a paltry 32,000 miles[86] covering an area about the same size as that of the US, so any large-scale effort to exploit shale oil and gas will require a massive outlay for pipelines and other petroleum transportation infrastructure to deliver these supplies to demand centers.

It's worth noting that companies such as ExxonMobil and Chevron are pouring money into countries such as Australia instead of China because of the Chinese government's over-involvement in gas and oil companies

85 EIA/ARI World Shale Gas and Shale Oil Resource Assessment, May 2013, p. 2
86 http://www.eia.gov/countries/cab.cfm?fips=ch

and tight control of the market. The state controls oil and gas pricing, keeping it artificially lower than what you see in the global market, which discourages investment in exploration. Beijing also recently implemented a new ad valorem (according to value) tax on oil and gas production, including unconventional resources.

The only way China is going to be able to meet its output goals is for the government to pour money into exploration and development and/or ease up on the price controls and taxes that undercut producers' ability to make any kind of meaningful return on their exploration investment.

Sometimes after my speeches, Chinese government reps warn me, "You need to make these comments more positive. We're not threatening you. It's just you need to be careful because you're a foreigner, an American on a visa, and you're basically bashing our country."

I try to explain that I'm just telling it like it is, but that's not the Chinese way.

The Chinese have no American-style entrepreneurial freedom. They have huge, state-controlled oil companies and none of the small independent oil and gas companies thriving in America with the freedom to take the risks that can result in the types of breakthroughs that generated our current energy renaissance.

China's options for energy security are limited if it wants keep a tight rein on its oil and gas imports. Beijing's strategy is to buy natural resources where it can while diversifying its oil and gas portfolio and to acquire knowledge about fracking by investing in companies in countries where it's being done.

China also faces plenty of competition almost anywhere it wants to play, and it faces something else that no amount of money can fix: After decades of being cut off from the rest of the world and emerging, still insular, only in the past thirty-five years or so, China remains somewhat awkward in terms of diplomacy. That has ratcheted up tensions with both its potential suppliers of energy and its competitors for that energy, especially in the Pacific Rim.

Poland's Energy Aspirations

Poland is another geopolitically important country that may be sitting on vast potential reserves of shale gas and oil.

Poland was the first major Soviet satellite to win its independence, and the Poles are eager to wean themselves from Russian energy. In my meetings with officials at PGNiG, the state-owned oil and gas company, it's clear they have aspirations to be a new gas hub for Europe, possibly linking up a pipeline that might one day stretch from the Mediterranean through Turkey and into Eastern Europe.

As for significant supplies from fracking, Poland may one day get there, but they've had setbacks and face some of the same infrastructure and other issues seen around the globe. One of the big issues in places like Poland is that the government has yet to decide on basic things like tax rates that influence how interested investors will be to risk capital on exploratory wells. Polish regulations, following trade policy measures of the European Union, prohibit the import of equipment and technology directly from the US. Approval for such imports is a red tape nightmare. Also, doubt has been cast on initial estimates of how much gas could be recovered by local geologists. In the aforementioned ARI assessment, a more stringent methodology was used, and Poland's estimated shale gas resource was reduced from 187 Tcf in 2011 to 148 Tcf in 2013.

Like so many other nations, Poland has no entrepreneurial oil and gas heritage. The state has a monopoly, and it's been so awkward to do business there that ExxonMobil backed out of a planned major development program. Although the official reason given was that two early wells failed to produce commercial quantities, ExxonMobil no doubt discovered what I've observed in Poland and elsewhere—you can drill a thousand wells and produce all the gas you want, but what good is it if there's no infrastructure to process, transport, or store it?

Also, Polish farms are very small. You need about five acres to set up shop on a drilling site, and that's a challenge on a ten-acre farm. Because the state controls what's in the ground, landowners are not entitled to royalties. They do get surface-damage payments, but that's not much incentive for a farmer who would lose half his productive land.

UK: Eurofracking's Bright Spot

The one bright spot for fracking in Europe is the United Kingdom, which initially balked at fracking but by the end of 2013 was moving toward adopting regulations that would allow the industry to fully develop.

Unlike most of its neighbors in Continental Europe, the British have a long history with the oil business, beginning in 1911 when Winston Churchill was appointed First Lord of the Admiralty. His mission was to modernize the Royal Navy, and one of the most significant changes he instituted was to switch from fueling ships with coal to burning oil, which produces about twice as much energy and thus allowed ships to move faster and travel farther.

The UK had plenty of coal but very little oil. Churchill's solution eventually became British Petroleum (BP), which had the rights to oil in Iran, then known as Persia. To protect its source of oil, Britain became more and more entangled in the Middle East, engineering regime changes and suppressing uprisings. All of the Western powers—France, Germany, the Netherlands, and the US—followed suit.

Over the decades, the struggle for control of oil in the Middle East and elsewhere played a key role in world politics, from World War II to the oil shocks of the 1970s, when new offshore technologies allowed the British to exploit and develop oil and gas deposits in the North Sea. From the early 1980s until the mid-2000s, the UK produced more oil and gas than it could use from its undersea fields.

Those were the good old days. Now, close to half the proven reserves have already been lifted, and production has been in a long slow decline since 1999. Though the UK is the largest oil producer and second largest gas producer in the European Union, it became a net importer of oil and gas in the mid-2000s thanks to production declines and long-term export contracts, and its production is expected to continue falling to about a third of its peak by 2020. Fracking offers a new source of energy that's relatively less expensive and carries less environmental risk—an attractive alternative to becoming ever more dependent on world oil markets.

According to a government-ordered report released at the end of 2013, more than half of the UK could be suitable for shale gas fracking and, in addition to easing reliance on imported fuel, the industry could generate an estimated 16,000 to 32,000 jobs. The report identified about 100,000 square kilometers (roughly the size of Virginia) available for drilling. The UK has all the advantages that so many other countries don't—plenty of water, good roads, open land, good infrastructure, and a lot of experience handling hydrocarbons.[87]

"We have a robust system of regulations and, provided companies have gone through due process, the map shows there is a huge amount of shale gas," Energy Minister Michael Fallon has said. The expected surge in energy development is already boosting wages of workers in the energy business. The government reported in January 2014 that oil and gas workers in the UK earn on average $130,000 a year, three times the national average, with yearly raises running nearly three times the rate of inflation.[88]

In December 2013, I was invited to give testimony at a Parliamentary committee hearing on fracking and was struck by how methodically the British are going about it. Unlike countries such as Poland, where the instinct was just to go drill some wells and see what happened, the British knew they needed a foundation and a framework. They understand that what's at stake is billions in investment and billions more in potential savings and economic stimulus. They will have the advantage of learning from the mistakes of others.

Still, fracking is just as hot a political potato there as it is here, and fears of a backlash have caused some of the larger oil companies to sit on the sidelines while smaller companies lead the way. That may be changing, though. In January 2014, French oil giant Total announced it is putting up about $50 million for a roughly 40% stake in shale gas licenses in the East Midlands.[89] The British Geological Survey, a research organization, said

---

87 https://www.gov.uk/government/uploads/system/uploads/attachment_data/file/273997/ DECC_SEA_Environmental_Report.pdf
88 http://www.bbc.com/news/business-25420552
89 http://oilprice.com/Energy/Energy-General/Total-Prepares-for-50m-Investment-in-British- Shale.html

last year that the region may contain 1,300 trillion cubic feet of gas. If just 10% of that gas could be produced, it would supply about forty-five years of UK gas consumption at current rates.[90]

After some fits and starts for UK fracking, Prime Minister David Cameron in late March 2014 insisted that fracking would get under way, with the first few wells expected to be completed in 2014. What prompted the ringing endorsement? Cameron said that Russia's annexation of Crimea from Ukraine should be a "wake-up call" over the need for Europe to become less reliant on Russian gas, reminding continental Europeans of the times Russia's squabbles with Ukraine had threatened Europe's supply of gas from Russia.[91]

Amid the furor over Russian President Vladimir Putin's Ukraine gambit, we have seen directly how the US fracking boom has become a key chess piece in the new "Great Game" between East and West: In late March 2014, President Barack Obama, while visiting with European leaders in Brussels, said the crisis in Ukraine underscores the need for the European Union to consider imports of shale gas from the US and to develop domestic resources to diversify supplies. He noted that his administration had already licensed the export of enough LNG to cover Europe's needs; however, those supplies are still some years away and are slated for the open market. At the time of publication, US and EU representatives were scheduled to meet again to discuss energy cooperation and ways to accelerate Europe's energy diversification.[92]

Fracking: Countermove to Russian Energy?

The Cold War may have ended more than two decades ago but East-West tension between Russia and Europe has remained a major irritant. Russia is one of the top three producers of oil and gas and Europe is its largest customer. From Belarus on the east to Spain in the west, countries from

---

90 https://www.gov.uk/government/uploads/system/uploads/attachment_data/file/226874/BGS_DECC_BowlandShaleGasReport_MAIN_REPORT.pdf

91 http://www.independent.co.uk/news/uk/politics/ukraine-crisis-david-cameron-pushes-for-fracking-to-lessen-reliance-on-russian-oil-9215431.html

92 http://www.bloomberg.com/news/2014-03-26/obama-pitches-shale-gas-to-europe-seeking-to-cut-imports.html

the Baltic Sea to the Atlantic buy about 80% of Russia's oil and gas output. The tension began almost as soon as the Soviet Union was dissolved in 1991 when Russia's big state-controlled gas company, Gazprom, repeatedly turned off the tap on the main pipeline to its newly independent neighbors such as Ukraine and the Baltic states over disputes about pricing and payments.

In 2006, with Europe having become dependent on Russian gas, Gazprom did it again, shutting off the pipeline that supplies Ukraine and also carries gas to the rest of Europe. That cutoff became a major international incident, with the US accusing Russia of intimidation and blackmail. Gazprom cut gas flows to Ukraine again in the winter of 2009 and Bulgaria, at the end of the pipeline, went without heating fuel for days until the two sides settled their differences in a new contract.

Russia has already taken a hit, if indirectly, from America's shale gas boom.

US imports of gas (as LNG) from Qatar peaked at 90 billion cubic feet in 2011; that trade evaporated a couple years later with the US shale gas production boom.[93] After the loss of the US market, Qatar trained its sights on Europe, helping to send Russia's share of the market there into a nosedive, from more than 40% in 2008 to about 25% more recently.[94]

In 2013, natural gas was at the center of a tug-of-war over Ukraine, which had been set to ally itself with the European Union. Instead, Russia put pressure on Ukraine's government to link its fortunes with Moscow by agreeing to invest $15 billion in Ukrainian debt and cutting the price of gas to Ukraine by a third. As much as anything, the Ukrainians' response to that overture was to depose their president in defiance of Moscow— hence the Crimean crisis.

To hear the Russians tell it, fracking is no big deal, just a passing fad. In May 2013, Gazprom Chief Executive Alexei Miller declared the US gas boom is "a bubble that will burst very soon. We are skeptical about shale gas. We don't see any risks [to us] at all."[95]

93 http://www.eia.gov/dnav/ng/hist/n9103qr2A.htm
94 http://bqdoha.com/2013/11/natural-gas-market-qatar-vs-russia
95 http://www.the-american-interest.com/blog/2013/05/01/us-shale-gas-boom-undermining-putins-gazprom/

That seems a bit disingenuous when you consider that Russia already was discounting gas to its European customers in 2013 because of increasing competition from energy supplies displaced by surging US shale gas production. Now Gazprom is starting to get a taste of its own medicine. Bulgaria forced Gazprom to cut its prices by 20%. "They can't bully us in the way they could before," a Bulgarian official said at the time. "We got the sense they need us more than we need them."[96]

Threatened with arbitration, Gazprom gave some of its European customers, including utility companies in Germany and Italy, $3 billion in backdated price rebates and new contract terms that allow them to pay spot prices instead of onerous contract prices. Meanwhile, former Soviet satellites in Eastern Europe are seeking new sources of gas and constructing new pipelines to be able to draw from other sources. And more competition is on its way.

It's no small irony that part of the lost Russian share of the European market was the result of some European customers switching from costly Russian gas to cheaper imported coal. And the source of that imported coal? The US, of all places, where coal exports are rapidly growing as US utilities and industries switch away from coal to take advantage of cheaper natural gas.[97]

Here's another irony: Despite its apparent disdain for America's fracking boom, it turns out Russia is no slouch itself when it comes to horizontal drilling and hydraulic fracturing. In recent years, Russia has been able to increase its oil production as it has expanded and enhanced its capabilities with horizontal drilling and multistage fracking.[98]

2040: US Still Number One for Shale

With all that said about the fracking boom spreading to other countries, it's pretty evident that the US will remain far and away the leader in using horizontal drilling and fracking for decades to come.

96 http://online.wsj.com/news/articles/SB10001424127887324240804578414912310902382
97 http://online.wsj.com/news/articles/SB10001424127887324240804578414912310902382
98 http://www.eia.gov/todayinenergy/detail.cfm?id=8350

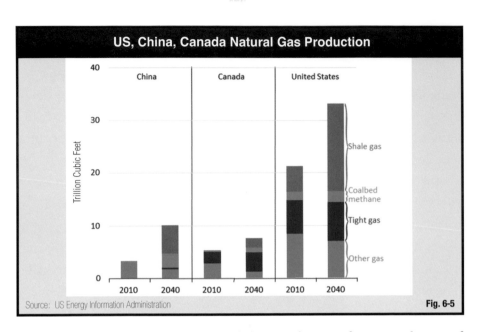

US, China, Canada Natural Gas Production

Source: US Energy Information Administration

Fig. 6-5

In assessing the most likely candidates to be significant producers of gas from shales and other unconventional sources, the US EIA settled on a comparison of Canada, China, and the US in 2010 and in 2040. Figure 6-5 shows the US still far ahead 25 years from now.

Aside from all the other considerations about resource potential and geology and infrastructure, the most compelling reason why the US fracking boom has been so successful and has completely transformed the outlook for energy both domestically and internationally is that it's purely homegrown American. No other nation can boast what we have that drives this boom: a healthy dovetailing of private landownership with mineral rights, a dynamic and entrepreneurial class of independent oil and gas companies plugged into capital markets, and an unsurpassed energy infrastructure.

However the energy boom plays out globally, the US is years ahead of the rest of the world in technology, production, and development of alternative energy sources, and we're making big strides in conservation. Our foreign policy may still call for us to act as the world's policeman from time to time, but at least it won't be because our energy security or our economy is at risk.

Now we only have to realize that mitigating what may—or may not—prove to be the biggest environmental risk of all—climate change—requires us to be clear-headed enough to recognize the potential solution that fracked natural gas offers. In short, if you want to arrest catastrophic climate change, you can't get there from here without an abundant supply of natural gas that comes courtesy of the fracking boom.

Bakken rig drilling in winter for Breitling Energy.

Chapter 7
Climate Change: A Game Changer?

Is Climate Change Real (and Does It Matter)?

I'm not about to suggest that I have any special knowledge or insights that qualify me to jump into the debate over climate change. I haven't entirely made up my mind on the subject. While a lot of scientific groups are in agreement on the likelihood of climate change, there are some legitimate scientists who question some of the claims, especially as they all seem to emanate from computer models that are frequently adjusted when the real data contradicts them. Or sometimes the data has been cherry-picked or even thrown out when it doesn't show the desired result, as we've seen in some recent scandals.

I get especially nervous when proponents of the theory that catastrophic climate change is looming dismiss anyone who questions their views as a denier, a heretic, or a tool of the fossil energy industry trying to sabotage their mission. Especially troublesome is the mantra, "The science is settled."

Okay, I'm no scientist. Far from it. But isn't "settled science" the very antithesis of the scientific method? Aren't scientists always supposed to be questioning the data, gathering more data, keeping an open mind that challenges the conventional wisdom? Some deride skeptics as "flat-Earthers," forgetting that at one time, the concept of a flat Earth was the settled science, as was the Earth being the center of the universe.

Some recent developments seem to be "unsettling" some of the climate science.

At this writing, the headlines are filled with the back-and-forth squabbling over the latest major report from the Intergovernmental Panel on Climate Change (IPCC).[99] That's the group that won the Nobel Peace Prize

---

Keystone XL: Key to Climate Change?

How did a pipeline become a flashpoint in the climate change debate?

One of the hottest controversies in the recent energy discussion has been the Keystone XL pipeline. This is a 1,179-mile extension of the existing Keystone pipeline that carries oil from Canadian oil sands projects and US Bakken shale oil deposits across the Great Plains to US refineries on the Gulf Coast. At first, the major objection to the project was that it crossed a huge and important freshwater aquifer and sensitive ecosystems in Nebraska. That objection was rendered moot with a route change. Then, the US State Department issued its 2013 report on the project's environmental effects, which found that the alternative to Keystone XL—shipping the oil via rail—actually increased the risk of oil spills and would create worse carbon emissions than the pipeline.

But Keystone XL has since become a rallying cry and symbol for climate change activists who have made it a litmus test for President Obama to demonstrate his commitment to fighting climate change. President Obama acknowledged as much himself, saying, "Allowing the Keystone pipeline to be built requires a finding that doing so would be in our nation's interest. And our national interest will be served only if this project does not significantly exacerbate the problem of carbon pollution. The net effects of the pipeline's impact on our climate will be absolutely critical to determining whether this project is allowed to go forward."[1]

So with the main focus shifted from local environmental impacts, Keystone XL has been even more heavily targeted by climate change activists, who

1 http://www.usnews.com/news/articles/2014/01/31/state-department-keystone-pipeline-report-finds-little-impact-on-carbon-emissions

---

99 http://www.nbcnews.com/science/environment/too-apocalyptic-experts-weigh-impact-climate-risk-report-n68151

for their last major report in 2007, along with Al Gore, for their work on alerting the public to the climate change threat. As it turned out, that 2007 report was riddled with errors, and the 2014 report seems to have walked back some of the earlier report's more alarming claims.[100] It seems that some viewed the report as too conservative, others as too alarmist.

apparently see its demise as another nail in the coffin of fossil fuels because, they think, it will halt Canadian oil sands development.

But that just won't happen. A subsequent State Department report found that the oil sands oil will likely move to market via rail if Keystone XL is disapproved. And in the past year a flurry of both rail and pipeline projects, some in the US but most in Canada, have been proposed as a Keystone XL alternative. Among those proposals is one that will result in shipping the oil sands crude west for possible tanker shipments to China. So even the prospect of Keystone XL being rejected ratchets up the carbon emissions/catastrophic oil spill risk even more. That latter State Department report also reiterated its finding that the project would not significantly affect climate change.[2]

In late February 2014, President Obama told a meeting of the nation's governors that he would make a decision in "a couple of months."

In fact, another eight federal agencies have until May 2014 to comment on the State report. After that, the Secretary of State will make a recommendation on the finding to President Obama, with the final decision his to make.

Reading the tea leaves at this writing, it looks like a toss-up as to whether the President will approve Keystone XL. Either way, his decision will be a lightning rod for criticism, depending on whether it's viewed as a job-killer or a climate-killer.

I see the project as one that enhances American energy security and actually poses less of a threat to the environment than the rail alternative. The latter point was brought home tragically in 2013, when a runaway, unattended train carrying Bakken Shale crude oil derailed and exploded, killing forty-seven residents and nearly destroying the village of Lac-Mégantic, Quebec.

2 http://www.washingtonpost.com/business/economy/state-to-release-keystones-final-environmental-impact-statement-friday/2014/01/31/3a9bb25c-8a83-11e3-a5bd-844629433ba3_story.html

100 http://online.wsj.com/news/articles/SB10001424052702303725404579460973643962840

Perhaps most damning about the 2014 report was the fact that one of the lead authors took his name off the report because it was too alarmist on assessing the risks from climate change.

And there are some, such as Matt Ridley, a member of the British House of Lords and vocal climate change skeptic, who maintain that climate change has actually done more good than harm so far and is likely to continue doing so for most of this century.[101]

He cites studies by Professor Richard Tol, a prominent economist with Sussex University, that concluded that climate change would be beneficial up to a warming of 2.2° C. as measured from 2009. But the latest "settled science" estimates of climate sensitivity are that we won't reach that level of warming until the end of the century—if at all.

By the way, Professor Tol also happens to be the IPCC report lead author who took his name off the report, not just because it was too alarmist but also, he claimed, because the report buried all information about the benefits of climate change.

What about the benefits of climate change? Ridley also notes the seldom-mentioned fact that more people die in winter than in summer and the winter death rate would be lessened by global warming. And while folks fret about climate change-induced drought afflicting crops and inducing mass starvation, other scientists have concluded that the greatest benefit from climate change comes from the CO_2 itself and its beneficial effect on plants. One scientist, using three decades worth of satellite images, has found that 31% of the world's vegetated areas have become greener while just 3% have become less green. Ridley notes that this translates into a 14% increase in productivity of ecosystems. Makes sense. Why else do commercial greenhouse operators pump CO_2 into their greenhouses? It's to increase the rate of plant growth.

101 http://www.spectator.co.uk/features/9057151/carry-on-warming/

Fracking: Not the Enemy on Climate Change

One of the more interesting arguments against fracking and the reduced energy prices that are resulting from all this activity is that there won't be a financial imperative to keep investing in renewable sources of energy that are noncompetitive on cost now but deemed by some to be essential to combat the threat of climate change.

But I would suggest that this argument contains its own built-in rebuttal of why we can't rely on renewable energy alone to slow down or halt global warming: It's not a cost-effective solution. Nor is it yet a workable solution, as we discussed in Chapter 1: Solar and wind power in particular, to name the two most prominent alternatives, require backup power capacity because their "fuels"—sunshine and wind—are intermittent and there is no cost-effective and environmentally friendly means for storing solar and wind energy.

We've seen the huge cost advantage that inexpensive gas from fracking is delivering to US industries, spawning a new American manufacturing renaissance. We've seen the energy cost savings Americans have been enjoying—even when the seemingly incessant string of polar vortex events pushed natural gas demand to a record high and set the stage for record gas storage injection at the end of the heating season. Spot natural gas prices did spike in January, but for the most part winter citygate gas prices remained below \$5/MMBtu. From 2005 through 2008, US citygate gas prices averaged nearly \$9/MMBtu, peaking at around \$12/MMBtu in summer 2008. Since then, it has averaged a little over \$5/MMBtu.

Like it or not, our economy—and the global economy—relies on affordable, on-demand, widely accessible energy to keep going. That means fossil energy. And if you want all that and a lower-carbon energy source, that means natural gas. This once-shunned byproduct of oil drilling has about half the CO_2 emissions of coal (Table 7-1).

This isn't a hypothetical benefit; the effects have been well-documented. Figure 7-1 shows the annual year-over-year change in CO_2 emissions in the US, and it's clear that the reversal of years of sequential gains in this

CO$_2$ EMISSIONS by FOSSIL FUEL	
Fuel Type	CO$_2$ Emissions/MMBtu
Coal (anthracite)	228.6
Coal (bituminous)	205.7
Coal (lignite)	215.4
Coal (subbituminous)	214.3
Diesel fuel & heating oil	161.3
Gasoline	157.2
Propane	139
Natural gas	117
Source: US Energy Information Institute	Table 7-1

main greenhouse gas corresponds with the surge in low-cost shale gas consumption by electric utilities and other industries.

On the strength of the evidence in favor of natural gas, many key government officials are recognizing the important contribution of natural gas to slowing the loading of greenhouse gas emissions into the atmosphere. Upon taking office in 2009, President Obama set a goal of the US cutting greenhouse gas emissions by 17% from 2005 by 2020, if all other major economies pledged to limit their emissions, too.[102]

US Energy Secretary Ernest Moniz in 2013 said that America was "about halfway" to the President's goal to cut greenhouse gas emissions and "about half of that is because of the substitution of natural gas for coal in the power sector."[103]

---

102 http://www.whitehouse.gov/sites/default/files/image/president27sclimateactionplan.pdf
103 http://www.politifact.com/truth-o-meter/statements/2013/aug/30/ernest-moniz/energy-secretary-moniz-says-us-halfway-greenhouse-/

| FOSSIL ENERGY EMISSION LEVELS | | | |
| Pounds/Billion Btu of Energy Input | | | |
Pollutant	Natural Gas	Oil	Coal
Carbon dioxide	117,000	164,000	208,000
Carbon monoxide	40	33	208
Nitrogen oxides	92	448	457
Sulfur dioxide	1	1,122	2,591
Particulates	7	84	2,744
Mercury	0.000	0.007	0.016
Source: US Energy Information Institute			Table 7-2

Even aside from the reduction in greenhouse gas emissions, natural gas comes out the winner in terms of power generation efficiency, which in turn makes for a 60%-plus reduction in CO_2 intensity, not to mention the other significant reductions in pollutants (Table 7-2).

That doesn't mean natural gas gets off scot-free on the greenhouse gas emissions side. Burning natural gas still emits some CO_2, but of greater concern are its methane emissions. Methane—the principal constituent of natural gas—is itself a potent greenhouse gas. A recent study, funded by the Cynthia and George Mitchell (yes, the "environmental fracker") Foundation concluded that actual emissions of methane from oil and gas industry operations were as much as 1.75 times what the US Environmental Protection Agency had been estimating. Fracking itself is a negligible source of methane emissions, the study concluded, and its assessment "still supports robust climate benefits from [natural gas] substitution for coal in the power sector."[104]

---

104 http://www.cgmf.org/blog-entry/92/Study-America's-natural-gas-system-is-leaky-and-in-need-of-a-fix.html

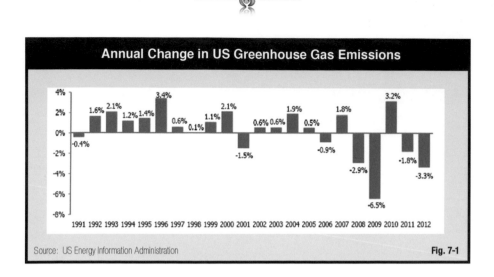

Annual Change in US Greenhouse Gas Emissions

Source: US Energy Information Administration

Fig. 7-1

Nevertheless, industry should make every effort to curb fugitive methane emissions from its operations. But it especially should move to minimize the practice of flaring natural gas. Note that I said minimize, not eliminate. There are safety reasons for sometimes flaring off natural gas. But those are brief, transitory events. Purposeful venting or flaring of natural gas for lack of market is wasteful and puts huge volumes of methane into the atmosphere. According to the International Energy Agency, flaring and venting of methane amounts to the emission of 1.1 billion metric tons of CO_2-equivalent per year. That equates to the entire annual natural gas production of Nigeria (which is probably the worst offender at this practice, by the way).

But we're an offender on this score, too. The flaring of about 1 billion cubic feet per day of natural gas, or about 30% of production, is happening right now in the Bakken Shale play, one in which my company is invested. We're flaring natural gas in certain areas of the Bakken because there's no way to capture the gas through a pipeline, and the Bakken oil can't be produced without flaring that gas. We're wasting that resource. I don't care if gas is worth 50 cents, it's still revenue if I can get the infrastructure in place to move the gas. It's been estimated that Bakken flaring wastes more than $1 billion per year in natural gas production value.

I'm happy to report that the industry in North Dakota has put together a task force to work with regulators to launch a campaign to capture nearly all of the natural gas being flared by 2020.[105] Solutions include accelerated construction of gas pipelines and gas processing facilities and stricter permitting requirements that call for operators to develop plans to capture natural gas when filing for permits.

I'm also encouraged to see that some companies are moving to utilize natural gas now being flared in their drilling and completion operations—with drilling rigs and fracking equipment fueled by LNG or compressed natural gas.

Whether or not you believe humans are responsible for climate change, we should do what we can to limit the amount of pollutants we put into the atmosphere. The pressure is on all over the world; it's become part of our daily dialogue and a factor in just about every business decision.

Conservation is a part of it, too. We've come a long way. Since 1978, per capita energy consumption has fallen more than 13% in the US.[106]

Technology is only going to make us more energy efficient in the future, and new sources of environmentally friendly energy sources will be discovered and exploited. But hydrocarbons are here to stay. Our entire infrastructure is built around this resource and we have it in abundance. There is no other choice when it's between continuing to import the energy we need versus perfecting the responsible discovery and production of it here at home. In any event, you can't make chemicals out of sunlight.

The decisions being made now about our energy future are going to be with us for many decades to come. It's up to everyone—the energy industry, government, activists, the public—to get it right the first time.

---

105 http://www.rigzone.com/news/oil_gas/a/132092/OG_Industry_Responds_to_Bakken_
    Gas_Flaring_Scrutiny/?pgNum=3
106 http://www.eia.gov/tools/faqs/faq.cfm?id=85&t=1

Drilling ahead at the Pumpkin Ridge #3 well for Breitling Energy.

Chapter 8:
The End Game

Precautionary Principles: For Humans Too

To me, one of the more convincing arguments to take action to prevent climate change is a pretty simple, commonsense one. It involves the precautionary principle.

Here's the definition of the term from Wikipedia: "The precautionary principle, or precautionary approach, states that if an action or policy has a suspected risk of causing harm to the public or to the environment, in the absence of scientific consensus that the action or policy is harmful, the burden of proof that it is not harmful falls on those taking an action."[107]

In other words: Better safe than sorry.

Although the concept has been around for eons—and is probably rooted in some basic human survival instinct—the precautionary principle really became imprinted on the public's mind with the earliest efforts to address environmental challenges at a global, intergovernmental level. It has since become enshrined in environmental law in some European countries.

And it has become a core tenet of the advocacy to take action to halt climate change.

I mentioned the actions and comments of Professor Richard Tol, one of the lead authors who took his name off the March IPCC report. Here's what another of the report's authors, Linda Mearns, a National Center for

107 http://en.wikipedia.org/wiki/Precautionary_principle

Atmospheric Research scientist who has studied the role of uncertainty in decision making, told NBC News: "The fact that we can't tell now exactly how much temperature will increase in Kansas City by 2050 is not a reason to do nothing. People make decisions under conditions of uncertainty all the time, and most people realize this—I hope."[108]

At first glance, that sounds sensible enough, but when you think about it, it reduces a complex problem to the absurd. I don't think anyone who's being reasonable in this debate claims that we don't need to do anything about climate change just because scientists can't offer a ridiculous level of precision in a climate forecast.

At the same time, those climate advocates should recognize the fact that a too-fast divergence from the energy systems that comprise the backbone of the global economy could wreck not only that global economy but put millions of human lives at risk of starvation, disease, and death.

In other words: The precautionary principle applies to humans, too, and not just to environmental values in the abstract. What I mean by that is that in taking action against climate change, we should emulate the doctor's Hippocratic Oath: First, do no harm. It should work both ways. But while I respect the views of the environmentalists—remember, I'm an environmentalist too—who worry about the damage from climate change to ecosystems and species and humanity, I believe that they should also consider the possibility of damage to the global economy that ultimately could be far worse for humanity than the planet warming by a couple of degrees.

Could experts like Professor Tol and Bjorn Lomborg (Director of Copenhagen Consensus Center, Adjunct Professor at Copenhagen Business School, and author of *The Skeptical Environmentalist*) be onto something? Maybe climate change is real but manageable. I can't imagine a system more complex and less understood than climate. When there's a weather disaster with extreme damage and loss of life, here come some climate advocates linking the disaster to climate change—perhaps obliquely, as in, "This sort of thing is happening with greater frequency because of climate

108 http://www.nbcnews.com/science/environment/too-apocalyptic-experts-weigh-impact-climate-risk-report-n68151

change." But when Al Gore is ridiculed for giving a talk on global warming in Boston during a record blizzard and subzero temperatures, climate advocates dismiss it, saying it's impossible to tie a specific weather event to climate change.

So now the climate advocates are hedging their bets a little, sort of like the last IPCC report: When a typhoon kills thousands in the Philippines, they might say, "Well, you can't tie a specific weather event to climate change, but you can expect more of this sort of thing because of climate change."

Remember Harry Truman's plea for a one-handed economist? You can't have it both ways.

The Environmental Threats We Know

The Skeptical Environmentalist author Bjorn Lomborg drew a lot of heat a couple of years ago for statements at the United Nations that centered on his claim that global warming is not the main environmental threat. He believes a greater priority than cutting carbon emissions in the near future is to allocate funds for research tackling longer-term challenges such as malnutrition and disease. Instead, we should find ways to adapt to short-term temperature increases because they are inevitable and manageable.

But what about the environmental threats that are killing millions of people every year now because they don't have access to affordable energy?

Case in point: One of the world's deadliest sources of pollution is particulate matter from soot, or PM2.5. This refers to particulate matter 2.5 microns or smaller, essentially microscopic dust particles from burning coal. PM2.5 particulates are so tiny that they penetrate deep into the lungs, get absorbed into the bloodstream, and cause cardiorespiratory disease.

This air pollution is killing more than 3 million people each year, mainly in the developing world, according to the Centre for Policy Studies (CPS), a UK think tank.[109] CPS published a recent study detailing this environmental horror while also debunking some of the

109 http://www.cps.org.uk/files/reports/original/131202135150-WhyEverySeriousEnvironme ntalistShouldFavourFracking.pdf

myths about the environmental threats posed by fracking shale gas and reached this conclusion:

> *Shale gas is urgently needed to address the greatest human-caused environmental disaster of our time, rising levels of air pollution, currently causing over three million deaths per year worldwide. At the same time it can dramatically slow the rate of global warming, and, as a bridging fuel, provide the time we need to develop truly sustainable non-carbon energy sources. The main dangers of shale gas can all be addressed by regulation to ensure that development is done using industry best practice, with heavy fines for malefactors. But why is shale gas needed in the developed world—a world that can afford to pay the premium for solar and wind? The fundamental reason is speed. Europe can develop shale gas far more rapidly than it can move to solar and wind, largely because of the low cost, the absence of an intermittency problem, and good existing gas infrastructure. To the extent that shale gas replaces coal, it will save hundreds of thousands of deaths each year, lives that will be lost if we choose the slower and more expensive transition to renewables.*

I'm especially intrigued by the term that the study authors used: "truly sustainable noncarbon energy sources." Whenever you hear about "sustainability" in reference to energy, it's invariably about transitioning to renewables as a driver of sustainability, a means to sustain environmental values.

But what about the economic sustainability of those energy sources? Shouldn't they sustain the impoverished, the hungry, and the sick whose lives would be bettered enormously by having access to affordable energy?

It isn't just particulates from coal. According to the United Nations, which declared 2012 to be the "International Year of Sustainable Energy for All," more than 3 billion people in developing countries rely on traditional biomass (wood, coal, etc.) for cooking and heating.[110]

---

110 http://www.un.org/en/events/sustainableenergyforall/

The World Health Organization recently estimated that over 4 million people die prematurely each year from illness attributable to household air pollution from cooking with solid fuels. More than 50% of premature deaths among children under five are due to pneumonia caused by soot from cooking with solid fuels indoors. [111]

Additionally, 1.5 billion people are without electricity, and even when it's available, millions of poor people are unable to pay for it.

Lack of affordable energy can also mean lack of adequate sanitation and lack of potable water, plus inadequate heating and cooling, so who knows how many more millions of people are dying from diseases caused by these parameters?

As I said, I'm an environmentalist but also a realist. Solar panels and windmills and energy efficiency and, yes, reduced greenhouse gas emissions are not going to do squat for these billions of poor, desperate people around the world. But maybe developing vast supplies of shale gas produced via fracking can displace enough coal to save thousands of lives each year right now. Whether it's transferring fracking technology to South Africa or Mauritania or providing gas via containerized small-scale LNG or CNG to poor villages in India or replacing coal consumption with natural gas in smog-choked Beijing, it could save lives now. If it also helps to reduce carbon emissions, well then that's a win-win.

Any way you slice it, we in the developed world have a moral obligation to assist the impoverished people in the developing world by promoting low-cost energy development as a means of ensuring a decent standard of living.

A Final Word

Fracking has been fraught with controversy, most of it based on misinformation like the claims made in Josh Fox's film *Gasland*. No matter how safe fracking is or can be, it's doubtful detractors will ever be satisfied.

---

111 http://www.who.int/mediacentre/factsheets/fs292/en/

Certainly my colleagues in the industry and I can do a better job not only in terms of embracing good environmental stewardship in our operations but also in communicating with and educating the public about what we're doing and why, and how we are in fact being good environmental stewards.

If I've been successful in achieving my goals in this book, you now have a feel for the big picture as well as the one in your backyard. The one in your backyard doesn't have to be the dangerous, dirty, toxic disaster some would have you imagine, and the big picture is that America has been given a second chance at security, prosperity, and global leadership. Let's make sure we get it right.

Index

Note: Page numbers ending in "f" refer to figures.
Page numbers ending in "t" refer to tables.

import facilities for, 108–111, 110t
importing, 108, 108f
liquids consumption, 121–123, 123f
Lomborg, Bjorn, 149, 150
Lynch, Michael, 118

M
Manhattan Project, 93
McAleer, Phelim, 81
McElhinney, Ann, 81
Mearns, Linda, 148
Mercury and Air Toxics Standards (MATS), 32
methane fields, 49f
methane hydrates, 38–40, 38f
Mexico, 108
Middle East
    oil fields in, 9
    oil production in, 7–8, 10f, 13
    oil reserves in, 7–8, 46, 54
    pipelines in, vi, 8–10, 10f
    volatility in, 13, 130
Migratory Bird Treaty Act, 26
Miller, Alexei, 133
misinformation, 46, 77–81, 152–153
Mitchell, Cynthia, 144
Mitchell, George P., 71–76, 144
Mitchell Energy & Development Corporation, 71, 76
Moniz, Ernest, 143
Motley Fool, 117

N
National Energy Technology Laboratory (NETL), 33, 34, 38–39, 43–45
New York Times, 81
Nigeria, 8, 54, 145
nuclear energy, 27–30, 28f

O
Obama, Barack, 15, 26–27, 132, 139–140, 143
Obama administration, 26–27
offshore resources, 13–14, 23, 50f, 75f, 130
oil. *See also* oil production
    access to, 13–15
    availability of, 37–55
    dependency on, 5–7, 6f
    peak oil estimates, 12–13, 12f
    peak oil theory, 9–13
Oil & Gas Journal, 43, 97, 100, 105

oil production in, 9–11, 11f, 13–15, 40–43, 42f, 97–107, 98f, 99f, 101f, 102f, 103f, 104f, 105f, 106f, 134–136
oil reserves in, 11–12, 11f, 97–99, 98f
oil resources in, 43–46, 44f, 51, 54, 104, 124–125, 124f, 125t
shale gas in, 51–55, 52t
shale oil in, 51–55, 52t
tar/oil sands in, 46
tight gas in, 48–51, 49f, 50f
Urban Land Institute, 75

V
Venezuela, 8, 54, 105, 125

W
Washington, George, 82
water pollution concerns, 22, 80, 85
whale oil, 3
wind power, 15, 22–27, 76, 92, 142, 151
Windfall Profits Tax Act, 73
Winger, Debra, 80
wood, burning, 1–3, 151
Woodlands Conference Series, 75
world energy consumption, 39–40, 40f, 122–124, 122f
world shale basins, 124–125, 124f

Y
Yamani, Ahmed Zaki, 39
Yergin, Daniel, 9, 11